Information Circular 9471

An Oral History Analysis of Mine Emergency Response

By Charles Vaught, Ph.D., CMSP, Michael J. Brnich, Jr., CMSP, and Launa G. Mallett, Ph.D.

U. S. DEPARTMENT OF HEALTH AND HUMAN SERVICES
Public Health Service
Centers for Disease Control and Prevention
National Institute for Occupational Safety and Health
Pittsburgh Research Laboratory
Pittsburgh, PA

April 2004

ORDERING INFORMATION

Copies of National Institute for Occupational Safety and Health (NIOSH)
documents and information
about occupational safety and health are available from

NIOSH–Publications Dissemination
4676 Columbia Parkway
Cincinnati, OH 45226-1998

FAX: 513-533-8573
Telephone: 1-800-35-NIOSH
(1-800-356-4674)
E-mail: pubstaft@cdc.gov
Web site: www.cdc.gov/niosh

This document is in the public domain and may be freely copied or reprinted.

Disclaimer: Mention of any company or product does not constitute endorsement by NIOSH.

DHHS (NIOSH) Publication No. 2004-145

CONTENTS

Page

Abstract	1
Acknowledgment	2
Introduction	3
Chapter 1.—Using narrative as a method to capture response veterans' knowledge	5
Chapter 2.—The narrators' involvement with mine emergency response	14
Chapter 3.—The decisions made during a mine emergency response	30
Chapter 4.–Some specific aspects of mine emergencies that affect responses	44
Chapter 5.—Sharing lessons learned	55
Chapter 6.—Summary and conclusions	66
Appendix A.—Surface organization of underground mine emergencies: interview guide	67
Appendix B.—Glossary of terms	72

AN ORAL HISTORY ANALYSIS OF MINE EMERGENCY RESPONSE

By Charles Vaught, Ph.D., CMSP,[1] Michael J. Brnich, Jr., CMSP,[2] and Launa G. Mallett, Ph.D.[1]

ABSTRACT

Beginning in 1991, scientists at the Pittsburgh Research Laboratory recorded interviews with individuals who are recognized as experts in mine emergency response. These 30 veterans related stories and observations from events experienced during as many as 47 years of response activities. Overall, the response veterans averaged 29 years of mine emergency response experience and 35 years of mining experience. Interviewees included representatives from mining companies, the United Mine Workers of America, and State and Federal agencies. Most of their comments dealt not so much with technical aspects of particular mine emergency responses, but rather the human side of more general topics, including preparedness, experience, people on-site, mine rescue teams, and decision-making. An analysis of their interviews provides an overview of lessons learned on-site at some of the largest mine disasters since the mid-1940s. This knowledge was gathered so that it could be provided to today's miners and to tomorrow's emergency response personnel. It is expected that the collective wisdom obtained can be used to help train new responders and guide those decisions that will have to be made on the scenes of future events.

[1]Sociologist.
[2]Safety and occupational health manager.
  Pittsburgh Research Laboratory, National Institute for Occupational Safety and Health, Pittsburgh, PA.

ACKNOWLEDGMENT

The authors gratefully acknowledge the contributions of James M. Peay (retired), who was instrumental in the conduct of this research.

INTRODUCTION

When an emergency occurs at a coal mine, there are four processes that come into play: information flow, sense-making, decision-making, and action. These processes are the way in which people generally impose a sense of order on a reality that the psychologist William James called a "buzzing confusion." The mental imposition of at least some degree of order on an emergency is necessary. Without it, responders would not be able to react in any meaningful way to the situation that they face. Because organization is essential to group action and since emergencies almost always call for concerted efforts, a primary goal of information exchange, sense-making, and decision-making is very likely to be the formation and use of a temporary organization.

It is in narratives about past events and their participation in them that respondents may reveal how they went about making sense and making decisions about how to organize themselves and others in the emergencies that they faced. Narrative has not usually been used in the study of organizational processes because it is not "scientific." There is little formal logic in narrative accounts, and they tend toward the common sense rather than causal abstraction [Czarniawska 1998]. However, it is through narrative that human beings come to possess most of what they know [Bruner 1986, 1990]. Whereas science is explanatory and depends on formal rules, narration is interpretative and grounded in the rules of everyday life. In all the areas of human experience where equations and logic are inadequate or need interpretation, therefore, it is narrative that serves to translate the unexpected into something intelligible.

This study portrays the processes of organization building and implementation from a historical perspective through an analysis of responders' narratives. Such an approach fits well with a recent development in organizational studies known as "knowledge management." Knowledge management applies to information flow, sense-making, and decision-making at both the individual and organizational levels. Generally, knowledge management involves some form of getting into people's heads and transferring what is there either to other people directly or to some type of repository that can be accessed when needed [Bassi et al. 1998]. The rise of knowledge management in organizational studies is a significant development because it is an implicit acceptance of the value of veteran workers' knowledge and supports the attempt to capture it in some structured and more usable way before it is taken out of the workplace forever.

The veteran informants for this study were interviewed about their experiences while working at mine emergencies. They were asked to discuss lessons that they had learned through experience. They were also asked to tell what they had learned that would cause them to handle similar situations differently and to tell about things they saw at past events that they would warn others not to do in the future. In response, the experts discussed a variety of things, but touched on some common topics. These included preparedness, experience, people on-site, mine rescue teams, and decision-making. This knowledge has been gathered and analyzed so that it can be provided to the next generation of miners in a concise and systematic manner.

This analysis is intended to reach as wide an audience as possible. Researchers and those who will be charged with training the future generation of emergency responders should find both the analytic and narrative portions helpful. Miners, managers, and enforcement personnel might be most interested in the narratives. For ease of use, chapters 2 through 5 address major questions from each section of the interview schedule. Chapter 2, for instance, addresses questions from section A regarding the respondents' involvement with mine emergency response. Thus, it is possible to skip the analysis and read just what the veterans had to say in response to each major question. Either way it is used, the collective wisdom of these veterans, presented in the chapters that follow, can be used to help train new responders and guide decisions that will have to be made on the scenes of mine emergencies in the future—a future in which emergency response will be a much less frequent, but nevertheless critical occurrence.

References

Bassi L, Cheny S, Lewis E [1998]. Trends in workplace learning: supply and demand in interesting times. Training & Dev *Nov*:51-73.

Bruner J [1986]. Actual minds, possible worlds. Cambridge, MA: Harvard University Press.

Bruner J [1990]. Acts of meaning. Cambridge, MA: Harvard University Press.

Czarniawska B [1998]. A narrative approach to organization studies. Thousand Oaks, CA: Sage Publications, p. 3.

CHAPTER 1.—USING NARRATIVE AS A METHOD TO CAPTURE RESPONSE VETERANS' KNOWLEDGE

This chapter pertains to three concepts that guided the authors' approach in this research: narrative, oral history, and knowledge management. Narrative and oral history refer to the systematic acquisition of knowledge in a spoken context. Knowledge management refers to the mechanisms used to store and use this knowledge once it is acquired. Although the concepts are fairly simple, there has been much academic discourse regarding the best theoretical explanations, most appropriate methodologies, and even the relative importance of such endeavors given the ascendance of electronic mediums.

The Greek historian Thucydides noted in *The History of the Peloponnesian War* [Crawley 1910] that he was either present himself at the events that he described or had heard of them from eyewitnesses whose reports he had checked as thoroughly as possible. Thucydides set a standard for the balanced use of individuals' narratives, and practitioners ever since have debated the proper way to collect and handle evidence obtained from such personal accounts [Hunt 1992]. Ultimately, however, when one cannot have been present at a historical event, personal narrative is an essential means of preserving the experience and passing it along to a younger generation. In the past 50 years, oral accounts have gained greater acceptance as a legitimate complement to the more conventional sources of scientific and historical record-keeping [Riessman 1993]. Social scientists and historians are able to conduct extensive interviews to uncover individual insights and opinions that are often omitted from written official documents. They have also noted that details of the planning and execution of many critical undertakings are now largely accomplished through electronic communication, with less information being committed to paper or put into permanent electronic databases. This further emphasizes the growing importance of oral accounts obtained through interviews, because the only place certain key organizational knowledge may exist is in someone's head.

Bock [2004] suggests that we look for key knowledge by asking: "What do we lose when key people leave?" or "What do we have to teach every new person?" The veteran responders interviewed for the present study were viewed as key people who had left, or were leaving, an enterprise in which much of the stock of knowledge resides in the oral tradition rather than on paper. In order to capture as much critical information as possible, researchers asked them to center their stories upon specific moments in the total progression of a mine emergency. Each account, therefore, follows themes: the first time I was involved in an emergency response, what it is like to arrive at a mine during an emergency, the hardest decision I ever had to make, etc. The researchers' task is to then make sense of the veterans' narratives in terms of each other and project research objectives. Weick [1995] described the most significant characteristics of sense-making from the research perspective:

> The answer is, something that preserves plausibility and coherence, something that is reasonable and memorable, something that embodies past experience and expectations, something that resonates with other people, something that can be constructed retrospectively but also can be used prospectively, something that captures both feeling and thought.

In other words, Weick considers sense-making to involve plausibility, coherence, and reasonableness even if the story must be filtered to make it acceptable and credible. It is the authors' intention to present information that offers insight, not to relate everything each person said about a particular topic. In this sense, the information is filtered and will be less comprehensive, "but, if the filtering is effective, more understandable" [Starbuck and Milliken 1988].

In relying on response veterans' stories to provide insight into past emergency response activities, the researchers draw upon the rich oral tradition in mining. They also, however, follow a recent trend toward the use of workers' stories in organization studies. The workplace became a legitimate domain for data gathering in the early 1980s [Martin 1982], and collected accounts were first treated scientifically and analytically [Czarniawska 1998]. More recently, according to Czarniawska, workplace stories have been presented in context with a minimum of academic intrusion. That is because the belief now is that their teaching value is greater if the stories are presented naturally rather than in a context-free "scientific" format.

Because organization is essential to group action, a primary goal of information exchange, sense-making, and decision-making is very likely to be the enhancement of organizational effectiveness. By current wisdom, an organization that can improve its performance the most and/or the fastest is the organization that succeeds. When it comes to organizational performance, according to Rubenstein-Montano et al. [2001], practitioners of knowledge management are already beginning to realize that worker knowledge is as fundamental to success as is technology. Given this more holistic approach, it is not surprising that some thinkers have situated knowledge management within sociotechnical or systems frameworks. The advantage to a sociotechnical systems perspective for knowledge management is that it delineates problems in their entirety [Hall 1999]. Thus, when knowledge management and systems thinking are combined, the resulting approach synthesizes environmental, technical, organizational, and personal perspectives. It also posits human judgment as a critical component of the decision-making process, recognizing that it is ingenuity that prevails in the unstructured conditions that characterize chaotic settings [Gorry and Scott Morton 1971].

The decision process, which depends upon explicit and/or tacit knowledge, begins with a problem. From a cognitive perspective, the person engaged in emergency decision-making about a problem is actively involved in a process characterized by a number of elements: (1) detection of the problem, (2) a definition or diagnosis, (3) consideration of available options, (4) choice of what is perceived to be the best option given recognized exigencies, and (5) execution of the decision based on what has gone before [Flathers et al. 1982; Baumann and Bourbonnais 1982]. At any moment in this process, it is possible for the individual to misinterpret elements, either because of environmental factors or because of his or her state of mind. When this happens, solution of the problem becomes more difficult and at some point will be rendered impossible.

Judgment theorists have focused on a limited number of variables that have a disproportionate impact on one's ability to achieve workable solutions to complex problems under limited time constraints:

1. The internal state [Hedge and Lawson 1979] is the sum of one's psychomotor skills, knowledge, attitudes, etc.

2. Uncertainty [Brecke 1982] is caused by faulty or incomplete information received from the external environment.

3. Stress [Biggs 1968; Jensen and Benel 1977] is generated both by the problem at hand and by any background problem that might exist.

4. Complexity, as it is used here, refers to the number of elements that must be attended to.

This model, by no means complete, nevertheless reflects the underlying demands on decision-makers in most life or death situations.[1] An emergency event imposes the necessity of dealing with an enormous quantity of sometimes faulty information in a very short timeframe.

Given the unstructured conditions found in an emergency, these problems are of the type that requires access to tacit knowledge [Polanyi 1966]. Tacit knowledge refers to the critical information people have in their heads that has not been written down. In recent years, organizations have directed a great deal of interest to tacit knowledge and to discovering ways in which it can be shared and used [Courtney 2001]. In Courtney's paradigm, the decision process consists not of leaping to some technically determined analysis, but of developing mental models by drawing upon the store of tacit knowledge residing in the organization. The key to managing tacit knowledge is an ability to start thinking in terms of shared learning [Addleson 2004]. Shared

---

[1] Steinbruner [1974] further discusses decision-making under complexity and uncertainty.

learning is the stuff upon which decisions are based, and these decisions lead to problem-solving activities. In order for an organization to be able to respond successfully, at least some of the responders must understand what is going on in the environment. By extension, the more rapidly that conditions change, the more critical shared learning becomes, and the more people are necessary to share in the learning process. In particular, workers in an emergency environment must be actively involved in developing, refining, and disseminating the group's shared learning while anticipating and responding to the changing conditions. Gray [2001] examined existing frameworks used to categorize knowledge management practices in order to explain why a problem-solving approach would be of value. He concluded that knowledge management is used precisely for the purpose of solving problems. Thus, knowledge management can be augmented by theories of judgment and decision-making that address problem solving as an activity that can be enhanced.

Theoretical debate about how people can be prepared to make decisions about the types of problems common to emergency response focuses on the importance of situation-specific knowledge versus mental process skills. Proponents of situation-specific knowledge point out that effective problem solving depends strongly on the nature and organization of knowledge available to individuals. Proponents of the role of mental process skills cite many studies that show it is possible to help people improve their abilities to think and make good choices in a wide variety of situations. In a mining context, both approaches have value and should be considered interrelated. For example, some miners may possess broad tacit knowledge about their work and their environment, yet not be able to use what they know in one particular situation in order to deal effectively with a similar situation having different content. In the same vein, some miners might have good judgment and decision-making skills, but know so little about the situation they are facing that their skills are of little use. It is when an individual can bring both knowledge and skills to bear on a problem that he or she will be truly effective.

Research Design

To this point, the terms "narrative" and "oral history" have been used almost interchangeably. Academicians, however, distinguish between the two, or, more accurately, show how one fits with the other. Narrative inquiry, in brief, is a way to better understand human experience through the use of people's accounts. "An account is the personal record of an event by the individual experiencing it, told from his point of view" [Brown and Sime 1981]. One of the most common applications of narrative accounting is the conduct of oral histories, which usually center upon historically significant experiences. Thus, an oral history may be seen as part of a larger scientific effort to understand human experience through the use of narration. Logic and experience suggest that there are various strategies for collecting an oral history account depending on the amount of control a researcher needs to exert over the interview process. If the research goal is to let narrators tell their stories however they wish, then an interviewer's focus will be on interaction. If the goal is information gathering, as in the present research, then a structured interview is appropriate. That way, the right questions will be asked and, although the narrator might digress, he or she will sooner or later be brought back to the task at hand. The focus of analysis is the qualitative data gleaned from such accounts.

There are still a few scientists who have not had experience with a qualitative strategy, which does not seem to fit their conventional notions of science, and profess difficulty understanding how the related activities can be legitimated. These scientists usually raise issues of reliability and validity when questioning the soundness of qualitative research. In general, reliability denotes the tendency of a measuring procedure to behave in a constant manner each time it is applied. The concept of validity is that a valid procedure measures what it is supposed to measure. For these scientists, the type of information gathering that depends on subjective responses has some major flaws. Yet, qualitative methods, which became virtually ignored in most disciplines following the rise of computer analysis and sophisticated statistical techniques, have had a phenomenal resurgence in the past decade [Miles and Huberman 1994]. Perhaps the chief reason for this renewed interest in, and use of, open-ended data has been the growing recognition that quantitative science leaves gaps in our attempts to answer "how" and "why" questions. In other words, the data never speaks for itself. No matter how well controlled the study is, a scientist who wishes to go beyond the immediate evidence and make statements about some population or universe must assume he or she "knows the relevant laws" [Campbell and Stanley 1966]. The strength of one's assumptions rests upon knowledge, experience, and creativity. If this seems to suggest that any type of science is only as good as its practitioners, such is at least partly true.

If there is a place in science for narrative analysis, what might that place be? Insofar as qualitative studies are concerned, they are usually thought of as exploratory or descriptive in nature. Appropriateness of a particular strategy, however, should be decided not by its nature but by the purpose for which it is being used [Kidd et al. 1996]. Yin [1984] listed three conditions that need to be considered before choosing a research strategy: (1) the type of research question being posed, (2) an investigator's extent of control over actual events, and (3) whether the events being focused on are current or historical. Questions that ask how or why certain contemporary events occur, but over which the researcher has no control, are particularly amenable to a qualitative analysis. Furthermore, a qualitative study, used as an explanatory mechanism, "provides evidence to show how both the rule, and its exceptions, operate" [Stoecker 1991].

The present research makes no effort to count something or measure a quantity. Instead, team members have tried to determine the behavior of responders in an emergency setting and then explain its variability in those instances where it is seen to exist. The pertinent question, in this case, is whether response behavior was studied by the researchers in a way that created a false reflection of it. Kirk and Miller [1990] suggested the proper response to that question: "it is incumbent on the scientific investigator to document his or her procedure." As in reports of quantitative research, the qualitative methodologist must make explicit the way in which the study was designed and carried out. In so doing, he or she guarantees that other scientists can determine whether or not the methodology is sound. They would then have an occasion to replicate the techniques, if appropriate, in other settings.

In qualitative studies, the fundamental tools used are a researcher's powers of observation or an ability to ask appropriate questions at the right time. A qualitative researcher often gains confidence in findings by using a structured instrument to examine an issue or variable of concern. The present narratives were expected to provide insights about the efficiency of particular emergency response efforts. The main data source of this study is a set of open-ended responses collected during interviews with responders who had worked mine emergencies at some time in their past. The main instrument used to gather these data was a structured interview guide (see appendix A).

Thirty individuals were interviewed in a private room and provided narratives concerning events in which they had been involved. The guide requested each individual to give an account of his personal experiences in the emergencies to which he had responded. Brown and Sime [1981] addressed the appropriateness of such an approach: "Fundamental to the philosophy of an account methodology is the recognition that people can and do comment on their experiences, and that these commentaries are acceptable as scientific data." If a person making an observation is skilled and his or her instruments properly constructed, then any subsequent conclusions ought to be considered accurate. In reporting these results, it is important to recount the methods that were used. Their appropriateness and proper use can then be evaluated by other researchers. Each of these scientists will ultimately decide if the instrument was constructed correctly and if the researchers were skilled in its use.

Nobody except one subject and two research scientists was permitted in the room during the interview. The rescue veteran was first asked for permission to tape record his account. All subjects agreed to be taped. A written schedule with a series of open-ended questions and related probes was used to guide every account. The exhibit below reproduces the major topic areas of this interview schedule. The entire instrument, with all questions and probes included, is contained in appendix A. Each interview began with an investigator reiterating that participation in this study was voluntary and that the responder had an option of not answering any particular question. After obtaining general demographic information, an interviewer next asked the individual the introductory question. Followup questions were then used so that specific details about each issue could be included. The sessions, which were 60 to 90 minutes long, ended when a researcher had asked all questions on the interview guide and a narrator did not have any additional information.

SURFACE ORGANIZATION OF UNDERGROUND MINE EMERGENCIES: INTERVIEW GUIDE

A. When I start the tape, I will ask you about your involvement with mine emergency response; when and why you become involved and things that stand out in your mind as you think back over your experiences.

Do you have any questions before we start?

[Tape on.] [Questions] [Tape Off.]

B. The next set of questions is about the decisions that are made during emergency responses.

Do you have any questions? Are you ready to continue?

[Tape on.] [Questions] [Tape off.]

C. During the next section of the interview, I will ask for details about specific aspects of emergency responses. Whenever you can, please answer with examples from your experience.

Do you have any questions? Are you ready to continue?

[Tape on.] [Questions] [Tape off.]

D. During this last set of questions, I would like you to think about all of your emergency response experience. I will be asking you about preparing future responders.

[Tape on.] [Questions] That is all the questions that I have. Thank you. [Tape off.]

The audiotapes were transcribed and stored on computer disks as text-based data. After the accounts were gathered, a comparative method of qualitative analysis was used [Glaser and Strauss 1967]. In the comparative method, a researcher develops as many categories as will clarify the problem. Next, he or she starts integrating categories and the properties that make them up, beginning to connect concepts with their indicators [Claster and Schwartz 1972]. After integrating categories and properties, the researcher is then ready to move toward simplicity and a broader scope [Glaser and Strauss 1967]. Over time, a theory of the event under investigation will emerge and be modified as more data get added. As the theory is streamlined, researchers are able to arrive at an assessment of how typical those occurrences that went into its construction are likely to be [Becker 1970]. The logic underlying this assessment is the same as that which supports probability: instead of adopting an either/or stance about the accuracy of particular assertions, one addresses the likelihood that his or her conclusions are correct. The magnitude of evidence from various data sources enables an observer to advance a particular conclusion with a greater or lesser degree of confidence. In the chapters that follow, the researchers will try to derive usable conclusions about some of the activities that may generally be observed in mine emergencies.

References

Addleson M [2004]. What is a learning organization? Fairfax, VA: George Mason University, Program on Social and Organization Learning. [http://psol.gmu.edu]. Date accessed: January 2004.

Baumann A, Bourbonnais F [1982]. Nursing decision-making in critical care areas. J Adv Nurs 7(5):435-446.

Becker H [1970]. Problems of inference and proof in participant observation. In: Filstead W, ed. Qualitative methodology. Chicago, IL: Markham Publishing Company, p. 194.

Biggs JB [1968]. Information and human learning. North Melbourne, Australia: Cassell Australia, Ltd.

Bock W [2004]. Knowledge management basics. [http://www.bockinfo.com/docs/kmbasics.htm]. Date accessed: January 2004.

Brecke F [1982]. Instructional design for aircrew judgment training. Aviat Space Environ Med 53(10):951-957.

Brown J, Sime J [1981]. A methodology for accounts. In: Brenner M, ed. Social method and social life. New York: Academic Press, p. 160.

Campbell DT, Stanley JC [1966]. Experimental and quasi-experimental designs for research. Chicago, IL: Rand McNally and Company, p. 17.

Claster D, Schwartz H [1972]. Strategies of participation in participant observation. Soc Methods Res 1(1).

Courtney J [2001]. Decision making and knowledge management in inquiring organizations: toward a new decision making paradigm for DSS. Decis Support Syst 31:17-38.

Crawley R, transl. [1910]. The history of the Peloponnesian war. [http://classics.mit.edu/Thucydides/pelopwar.html]. Date accessed: January 2004.

Czarniawska B [1998]. A narrative approach to organization studies. Thousand Oaks, CA: Sage Publications.

Flathers GW, Giffin WC, Rockwell TH [1982]. A study of decision-making behavior of pilots deviating from a planned flight. Aviat Space Environ Med 53(10):958-963.

Glaser B, Strauss A [1967]. The discovery of grounded theory. Chicago, IL: Aldine Publishing Company, pp. 101, 111.

Gorry GA, Scott Morton MS [1971]. A framework for management information systems. Sloan Manag Rev *13*:98-110.

Gray P [2001]. A problem solving perspective on knowledge management practices. Decis Support Syst *31*:87-102.

Hall MLW [1999]. Systems thinking and human values: towards understanding performance in organizations. In: Parra-Luna F, ed. The performance of social systems: perspectives and problems. Hingham, MA: Kluwer Academic Publishers.

Hedge A, Lawson BR [1979]. Creative thinking. In: Singleton WT, ed. The study of real skills. Vol. 2: Compliance and excellence. Baltimore, MD: University Park Press, pp. 280-305.

Hunt RA [1992]. Foreword. In: Everett SE. Oral history techniques and procedures. [http://www.army.mil/cmh-pg/books/oral.htm]. Date accessed: January 2004.

Jensen RS, Benel RA [1977]. Judgment evaluation and instruction in civil pilot training. Report No. FAA-RD-78-24. Savoy, IL: University of Illinois Aviation Research Laboratory. NTIS No. AD A057 440.

Kidd P, Scharf T, Veazie M [1996]. Linking stress and injury in the farming environment: a secondary analysis of qualitative data. Health Educ Q *23*(2):224-237.

Kirk J, Miller ML [1990]. Reliability and validity in qualitative research. Beverly Hills, CA: Sage Publications, p. 72.

Martin J [1982]. Stories and scripts in organizational settings. In: Hastrof A, Isen A, eds. Cognitive social psychology. New York, NY: North Holland-Elsevier, pp. 165-194.

Miles MB, Huberman AM [1994]. Qualitative data analysis: an expanded sourcebook. 2nd ed. Thousand Oaks, CA: Sage Publications.

Polanyi M [1966]. The tacit dimension. Garden City, NY: Doubleday and Company.

Riessman CK [1993]. Narrative analysis. Newbury Park, CA: Sage Publications.

Rubenstein-Montano B, Liebowitz J, Buchwalter J, McCaw D, Newman B, Rebeck K, et al. [2001]. A systems thinking framework for knowledge management. Decis Support Syst *31*:5-16.

Starbuck W, Milliken F [1988]. Executives' perceptual filters: what they notice and how they make sense. In: Hambrick D, ed. The executive effect: concepts and methods for studying top managers. Greenwich, CT: JAI Press, p. 41.

Steinbruner J [1974]. The cybernetic theory of decision: new dimensions of political analysis. Princeton, NJ: Princeton University Press, pp. 47-87.

Stoecker R [1991]. Evaluating and rethinking the case study. Sociol Rev *39*(1):94.

Weick K [1995]. Sensemaking in organizations. Thousand Oaks, CA: Sage Publications, pp. 60-61.

Yin RK [1984]. Case study research: design and methods. Beverly Hills, CA: Sage Publications, p. 16.

CHAPTER 2.—THE NARRATORS' INVOLVEMENT WITH MINE EMERGENCY RESPONSE

Mine rescue work is voluntary. Although there are many individual reasons that a person might join a volunteer organization, they can be placed under three overarching elements: (1) instrumental, (2) affective, and (3) altruistic. Instrumentality usually refers to the rational use of one's relationships with other people to achieve personal goals. An example might be a person who joins a mine rescue team in order to enhance his or her skill set and move up in the organization.

> SPEAKER: It was something I was interested in and, at that time, there was a lot of fires in the mines and jobs were kinda hard to get, and if you had the mine rescue experience and first aid experience, then you could just about get a job in one of the mines here, that and with mining papers. So I figured...my education like it was I had to do something to kinda bridge the gap.

> SPEAKER: My dad started me, talking about it, and I really done it to make more money. That's exactly why I did it.

Affect is a condition in which one's relationships with others are valued in and of themselves. A person might join a mine rescue team in order to share in the camaraderie such an experience offers and not be too concerned with advancement or other "rational" goals.

> SPEAKER: I trusted, definitely trusted, the people that I was actually working [with] at the time and they also trusted me.

> SPEAKER: The training program and the equipment that you use and the team membership camaraderie that was established at that time, I think that pretty well drove me on the path to where I am today, is that experience.

Altruism describes a rationale for conduct in which the good of others is the ultimate end for one's actions. In theories of ethics, altruism is the opposite of self-interest. A person motivated by altruism might join a mine rescue team in order to "give something back" to the organization or the community whether or not advancement was a possibility and perhaps despite having a low need to bond with his or her buddies.

> SPEAKER: I was always interested in mine rescue and Peabody 40, which is in the southeast part of Illinois, near Harrisburg, was low coal.

It ranged from 40 to 48 inches. And the State rescue teams at that time were using the McCaa, which is a backpack apparatus, and it could not be used in low coal effectively. So Peabody developed their own team and we were trained with a McCaa, but we were also trained with the all-purpose gas mask, and the Chemox apparatus, which was a breast apparatus that you could wear in low coal. So they wanted their own team which could respond to an emergency in the mine. Also, you had to be conditioned for low coal and most of the teams, the State teams, who worked high coal could not come down there and function very well as a rescue team with the backpacks on. This was in 1950 when I went on that team.

Obviously, as can be seen even in the examples above, an individual is more likely to have mixed motives about becoming involved in rescue and recovery work, so that his or her explanation for volunteering is less clear-cut than it might be otherwise.

SPEAKER: I feel like I have excelled in mine safety, and I hope I've left some type of mark. The superintendent who hired me...had one eye, and most people referred to him as being able to see more with one eye than most people could with two, but he told me that studies have been made that the more you learned about mining and the more training you had, that the less potential you had to have an accident. And he also told me that the more I knew, the better opportunity I had to keep in work, because in the coal industry you had the ups and downs and the layoffs and so forth, and from the first day I made an effort to get involved in as much as I could. I was immediately requested to go on a first-aid team and today's miners, you see them being paid time-and-a-half, or extra work for being on a first-aid team or mine rescue. In those days, if you was on it, you had to do it on your own time. And I put my own time in, and I was on a competition first-aid team, and then later on, I think it was 2 to 3 years later, I got on the mine rescue team, and I got deeply pretty involved in mine rescue after that.

Regardless of their motives for volunteering initially, however, the reasons individuals remain in mine rescue are more likely to be found in the rewards of their immediate work experience [Pearce 1983]. In other words, there must be something about the emergency situations themselves that provide incentives to continue in the activity. Researchers tried to uncover some of these motivations by asking about the first time the responder had witnessed an emergency situation at a mine.

SPEAKER: I was at a mine in 1978 and I was at one end of the mine performing a regular inspection and I received a call. I hadn't been on the section very long, and I received a call that my ride was ready to take me out of the mine. And I said, "I just got here; I'm not ready to go." I noticed people were real antsy, jittery, and I continued on with inspection work, then one of the people that I knew there grabbed me by the arm and said, "your ride is ready," and I knew then that he had something to tell me that he didn't want to repeat in the whole group of people. So we got to a place where he could talk. And he said, "We've had a mine accident in the other side of the mine and we think we have three people killed—covered up in a roof fall." He said, "They want you to go to that area right quick." And I said, "Okay, we'll leave." So we got on a conveyance that took us to the other part of the mine, which I'd say was probably 4 miles away underground. And we got to the accident site, and by the time we had gotten to the accident site a person had already been taken out that had been killed. And there was a person trapped under a rock fall, and we could talk to him and everything was in a real confusion.

I knew that I should start taking notes and start documenting all the events that took place, and I played a role as I just sat back and monitored what was going on. And as [I] watched the recovery effort trying to get the miner out, [there was] a lot of confusion going on and I got the head guy that was in charge and I took him aside and I told him about a lot of the confusion that was taking place, and I advised him that, you know, you might do better if you delegate some of the responsibilities out or organize it and also start working on backup efforts, that this thing may go on for an extended period of time, that he needs to be thinking about new fresh people, supplies, and taking care of what might take place on the other shifts and things, and then they started that. I can't tell you how it happened, but it ended up that I just quit taking notes and became involved directly in the recovery effort, and it ended up being divided up into two teams. And one team attacked, again we could talk to the gentleman, we couldn't see him, we knew fairly close to where he was, but we really couldn't identify him. They ended up digging a trench to this gentleman.

He had been inside of the cab of the continuous miner, and when the rocks started to fall, the rock generally fell in one large chunk. And it was approximately from 0, from a knife point, up to 7 feet thick and 20 by 20 feet large. And when the rock fall fell, it mashed the cab or the canopy of the continuous miner, but he had exited the cab area and tried to run and what it did, he made it to the rear bumper of the continuous miner and he was laying, well, the frame of the continuous miner supported the major part of the rock. And it had broken right at the end of the boom of the continuous miner and he was laying in an area that supported the major part of the weight but that rock, the break on the rock at the end of the boom, had crushed his hat and had his left hand trapped under the rock fall. So he was mashed, and the weight of the rock had pushed the continuous miner down into the soft floor structure and he was trapped there in a very restrictive position—he couldn't even get a full breath of air. And we had two groups. One was trying to break the rock that was closest to him with hammers and stuff like that—just chip the rock. They already had the area resupported with timbers to keep more roof from coming in, protecting the guy. And we were on the other side of the continuous miner and, by luck, it happened to be that the floor was a very loose material and [I] took a wedge that was used to tighten up timbers and that's what I used to scrape and I actually scraped a trough, a little indentation in the floor bottom, and I crawled. As I belly-crawled forward, I would scrape it, then I would kick, scrape the material back to my knees and then take my knees and feet and kick the material behind me. I was like a mole. And then the people who were behind me took that material and got it out of the way by the same method as we ended up going in. And I ended up creating a, again, this mole way, if you want, around the end of the boom and I got to this individual and I was able to talk to him and he was facing away from me with his right hand trapped back along the side of his body and his left hand out over his head like this with the roof fall trapping his hand.

I got to where I could see and I could hear him talking, you know, and he would respond to me. I remember talking constantly and what I was trying to do was keep him from going into shock or from listening to what was going on. His name was Ollie Bryant and—all kinds of things—time stands still when you're involved in it. And I remember talking to him and I was trying to figure out what am I going to do with this guy. His hand is trapped and am I going to have to cut his hand off to get him out from under this rock? And the first thing I did, I took my right hand and I started pulling, clawing material and I reached in under his neck to right where his mouth was and just raked material out of that way, and I felt a gooey mess and I thought he had internal injuries and he was bleeding and that is what had created this gooey mess, and when I pulled it back and looked, it was tobacco juice. The guy still had tobacco in his mouth and had not gotten rid of it. He was still doing this talking. So, you know, I was really relieved at that. Then I wondered what am I going to do about his hand being trapped. I just kept clawing material out from under and I got his hand free, but still, because of the way he was, he was trapped and couldn't get free. Then, I clawed loose material [and] made a trough right along beside the length of his body. [He] had a new belt, a big, wide miner's belt, leather belt, and I said, "I'm gonna have to cut this belt off of you in order to get you out of here." And he said, "Don't cut my belt off." And I said, "I've got to." I said, "Why don't you want me to cut that belt off?" I said, "The superintendent will give you a new belt." He said, "The hell with that superintendent, I made that belt." He didn't want it cut, but I cut it off of him anyway. And it had his self-rescuer and his miner's light on it. And I took my knife and cut that belt off of him and he had bib overalls on. And I told him I'm gonna count to three and I want you to exhale all you can 'cause I'm gonna yank you out from where you are. And I said, "One, two, three." And he exhaled out and I just grabbed his bib overalls and I just pulled him over into that little indentation I created for him. And he said, "Boy, that feels good." And I said, "Come on. We're getting out of here." And he said, "Just a minute, I got to get a breath of air." He said, "I haven't been able to breathe." I said, "Get you one and we're gone." And I got hold of his bib overalls, about like this, and I told the people outby me, I said, "Pull my legs." And it ended up being there were three of us. They grabbed, it was like a train, they grabbed my legs and I grabbed his bib overalls and they pulled us, snaked us out of that area, and I had more scrapes and scratches on me than he did. The minute we came out from under where the rock had fallen, he sat up and he said, "Boy, you all look good." And I was tending all the scrapes and stuff 'cause when they pulled me out, it just scraped me all real good up to here. And you know, they brought a scoop up there and, boy, we were like heroes then. They escorted us out of the mine, and [we] all went to the surface.

By the time we got to the surface, the company president was out there and all kinds of press. Boy, they were calling in and all kinds of really crazy reports were coming in that a president was sitting out there. And he was just out of control really, he didn't know what to do and I told him, I said, he had a secretary that I knew 'cause it was one of the mine rescue team member's wife was secretary there, and I knew her and I told him, I says, "What you need to do is call her, give her a general statement to make, to report." And I said, "Let her handle all that, we can continue with whatever business you have to take care of here."

But that was the event. There has always been an oddity about it because the next day for the investigation I was able to go down, but I've never got asked anything about the roof fall. I always, since 1978, was never asked in any official capacity about that event. I never had to sit down and really, really, go through it. I got to go down, [but] I didn't get to meet all those people that took part in it. For the next 3 years as I went back to my regular inspection activities, I would run across individuals, some of them even changed jobs and went to other mines that remembered, and I was able to meet a lot of that crew and found out a lot of the behind-the-scenes things that took place and stuff like that. But it was just a really odd experience to go through.

The account given here fits well with the general activity model of participation [Smith and Macauley 1980], which holds that the more one participates in a socially approved activity, the more likely he or she is to remain a participant. This is especially true if the volunteer is emotionally stable, has a strong ego, and is assertive. The narrator of this account was able to put aside his official role and assert himself, not only with the trapped miner, but with the company president as well. He was able to problem solve throughout the event, received ego gratification by being treated as a hero, and social recognition for several years thereafter from those who had been at the scene. All of these factors together undoubtedly played a part in his continuing mine rescue involvement.

Smith [1994] noted that perceived group effectiveness and positive outcomes are two key determinants in volunteer retention. The following account of a rescue in a U.K. conventional longwall operation is a case in point.

SPEAKER: I'd already taken the mine rescue training. I was part of the team, and it was my first experience as a mine rescue team member in an emergency situation. What had happened, there was a methane outburst at the neighboring colliery. And the full-time brigadesmen had already been called in, and we were called in as volunteers from the neighboring mine to assist in that rescue attempt. I was on the third team underground, and this was a conventional longwall. The methane outburst was to the magnitude where it completely stopped the ventilation.[1] There was that much methane coming out of the surrounding strata. The assignment that we were given was to go into the tailgate or the return airway and travel that return airway, taking samples and looking for any survivors. The intake airway had been damaged to the point where some ground support had to be put in, and there were teams working on achieving that. When we entered that return airway, the first samples we took gave us a 56. If my memory serves me right, about 56% methane and the rest was associated gases. So we knew we were in a real inert atmosphere that wouldn't support life. And traveling in methane at that percentages were very, very unnerving. The travel that we had to undertake was approximately 1,200 feet. So, in distances, it wasn't too far. There was no smoke. Obviously, the visibility was really good.

---

[1]Ventilation is an essential factor in safety, health, and working efficiency. It is also necessary to dilute and remove noxious or flammable gases and to abate problems such as dust and high temperatures.

I was the No. 2 man on the team right behind the captain. We were carrying two Survivair rescue packages. We was carrying an old Army-style, foldup canvas stretcher and first aid equipment, along with mine gas detecting equipment. I remember the first feeling that I had in relationship to the life support system [possibly failing] came into my mind around 15 minutes after we entered that return airway. And that was a very, very uneasy feeling at the time, knowing that if that apparatus failed, your team members would have to be pretty quick to get you onto one of the rescue apparatus that we were carrying. After about half an hour travel, we, the captain stopped. We were wearing the ProtoMark IV, which you have to wear a mouthpiece, you have no face mask. So communications were all done on a little writing tablet the captain carried. And the captain stopped and he halted the team and he did a team check. When he turned around and looked up towards the longwall face, he saw light and I saw light, a cap lamp. And he wrote on his little pad and he said, "Did you see that light move?" And I just shook my head and said, "No." So he stopped and he wrote on there, "Let's take a little bit of time and watch and see." And sure enough that light did move.

Now, we had a situation where we were working in a high concentration of methane. And here, in the distance up the roadway, we could see this light moving. The thought that had come into the team captain's mind, and my mind, was the fact that we would now, if an atmosphere further ahead is supporting life, have to come down through the exploding range of the methane. We discussed that by writing the little notes, and the captain said, "Well, should we continue or should we go back and report this? We've still got another hour and 15 minutes left on a 2-hour oxygen tank that we're on." And that was the assignment, to go as far up to the longwall as we could to see if we could rescue anybody that may be alive. Again, that was another experience that is very hard and very difficult to explain, insomuch that nothing could support life. But here again, you see this light moving. So we decided that we were going to continue. A sense of urgency was there now. The uneasiness was there, not knowing what to expect, and we did continue.

We arrived at the, what we call the "ripping lip." It's a 3-foot seam of coal, and the coal would be mined out underneath the gate or the roadway, and there was a compressed air line that run in to drive the undercutting machine. There was two individuals, one was a gentleman of, in his early 60s I would imagine, and the other young man was 16 or 17 years of age. And their main job was to take supplies off this drift-in roadway to the longwall face for the miners to use—timbers, roof supports, and whatever. Now, what this elderly miner had done was he took some plastic, when he heard the bump and he heard people up the longwall hollering and screaming he knew there was something wrong, so he took the clear plastic and threw it over the top of both him and his coworker, and took his knife and cut the compressed air hose that was going to the cutting machine. So now, they were more or less sitting in this little bubble and surviving off the compressed air that was down in the area.

Management's decision not to turn off the compressed air was made unaware to me at the time. The mine manager made that decision that the compressed air that runs down that longwall, we have to maintain, because somebody may be using that to survive on. They knew that the compressors weren't dumping enough loading. So they knew this compressed air was leaking or going someplace. On arriving at the scene, trying to communicate with these individuals was very, very difficult. The younger man was in some sense of shock, some extreme anxiety. The older miner was calming him down and saying, "We'll be fine." The captain wrote underneath, "We have two breathing apparatus, and we're going to put them underneath, and we want you to put them on. Here is how we do it." And he wrote them a step-by-step explanation on how to do that. One of the apparatus unfortunately didn't function. The one that the elderly miner put on wouldn't function. There was some problem with it. The other one did function. So what we did was, we explained to the elderly guy, "Stay there. We will get another team to come in to get you. You're fine. The compressed air has been maintained." And he just said, "That's fine. I'm fine. Don't worry about it." So we took out the young victim and we got him out to the area, and another team went in and brought the elderly guy out. That was my first experience in mine rescue, and it is a little difficult to try and explain to people the thoughts that were going through your mind at the time. But that experience was so enlightening, so fulfilling, that you would be able to take that training and then apply it and rescue somebody who needed your help.

Notably, most of the accounts of initial rescue activities contain themes of problem-solving, fortuitous occurrences, and the expectation that an individual's efforts may bring good results. In short, while money might be an initial prime motivator for some of those who engage in mine rescue work, it is more likely the "enlightening," "fulfilling" application of one's training to help a fellow miner that keeps a person engaged.

It is unfortunate that in mine rescue the notion of achieving good results must extend to the safe recovery of bodies at least as often as to the rescue of live workers or to sealing the mine as often as extinguishing a fire. Most rescue veterans were attuned to the fact that whether a response was likely to have desirable outcomes or not, they should "hope for the best but prepare for the worst." In his book *Sensemaking in Organizations*, Weick [1995] noted that people use strategy as a framework to help them minimize surprises. In instances where strategy is not well laid out, the "worst" has not been prepared for, and surprises may abound.

SPEAKER: The explosion at West Frankfort was in December of '51, I believe. At that time, I wasn't the team captain, [but had] been on the team since the Centralia explosion.[2] But the team captain didn't know whether he could handle it or not; he'd worked the mine. So before we went below, they made me team captain. And our team found about 67 of the 119 people that was in there. At that time, they would lay out what we were to do, or where we was supposed to be, and where we might find some people. It wasn't that well organized because some of the teams that come in to relieve us went back out with us at times. Our team and the Du Quoin team spent an average of 12 hours a day there for a week. That's inside doing the rescue work. The teams that we worked with were pretty good rescue people. What we tried to do is keep one team at the fresh air base and stay there till this other team came. Now at one time they come in and wanted us to go spray some bodies in a panel off to the right. This is about the third or fourth day and we wouldn't go till we saw a team come back. And one of the men that went down to spray them took two guys, and they got down. So as soon as we saw a team, counted the lights coming back, we did go and get these two guys that got down over there. But with that, they didn't come back anymore, the company took 'em and told 'em not to come back. But you have to have faith in the people that you're working with.

It was kinda disorganized, which you can see why. There was 119 people missing. And our team was a mixed team that got the last man out of there that was found alive. And I drew the map of the place where we found these nine dead ones and one live one. And that live one was on top of a rock fall, and these other nine around him was dead. But he said, "Here. Here. I knew you'd come." And then there was one big red-headed fellow sitting next to the rib looking right at me. I spoke to him just like he was [alive]. But he was [dead]. And we had to take him out around a bunch of cribs and stuff and rigor mortis had set in. We had to unload him a couple of times to get him out of there, even. There was so many people coming; just sightseers. They were supposed to come in and pack the bodies out. Well, they might come in and you'd see them maybe coming to the first body—they might bring him out [or] they might turn around and go back. Well, they didn't have that too well organized.

SPEAKER: We got a call—we had won several contests as a mine rescue team—so when Cargill had their fire in the shaft, it was recommended that the team from Dekoven and the Island Creek team from Madisonville would come down. They had an airplane sitting at Madisonville, and they called us from underground. We went to Madisonville and loaded all our equipment on, and they flew us to

---

[2]Centralia No. 5 Mine, Centralia, IL, March 25, 1947; 111 killed.

Morgan City, Louisiana. They had two helicopters to pick up both teams and take them out to the Island. When we got there, everything in the shaft had burned out, [including] the two hoists, and it was 1,285 feet deep. There was no way to get in—no way to get out. The two teams that was there had never experienced using a shaft, we had all been coal miners—we had gone in and out a slope, so really, no one knew what to do or how to do it. But we all got our heads together and decided to get a new rope on the hoist [because] the old rope was gone. They finally got a rope on, and we first made a square platform to put on the rope to go in and out, decided to try it, and it got down a few feet and hung up, so we knew that wouldn't work. Then we ended up making use of a piece of pipe, 36 inches in diameter, on the rope to let people up and down. And that's the way we got in and out of the mine.

Of course, the rules say that you won't travel more than a thousand feet without fresh air. And its 1,285 feet to the bottom of the shaft, so what do you do? Do you stop? Not really. So we made several trips before we could ever get to the bottom, because everything was burned out of it. And once we got to the bottom, an extension going down was filled with water. It's about 35 feet deep. And we had to manage to get the bucket from the center of the shaft over to get out. So everything that you try to do seemed as though it was working against you. But we finally did get into the mine, and we explored the entire mine using McCaa machines and gas masks. And once we located the 21 people, they was scattered all over the mine.

We did not feel that it was a good idea to have so many people underground [to recover bodies] with no fresh air. Carbon dioxide was still there; the air was not that good. So we wanted the oil drilling companies to come in and put a 6-inch line down 1,285 feet, and I happened to be one of them that constructed a fresh air base. We cut everything, put it together outside, and took it apart, we marked everything, and we carried that along with the people down the shaft and over a short distance from the bottom. Put it back together, covered it in plastic. In the meantime, they dropped the 6-inch line in, put an air compressor on the outside, and that's the way we had a fresh air base at the bottom of the shaft. At that particular time, I happened to be the one that they felt could put it together. I did it outside, then I went underground and I worked 6 hours using gas masks underground, which again you're not supposed to do—when they brought me up in the bucket I couldn't get out of it, I was that exhausted. So that's the kind of things that you actually do when there's a need for it. It took us 7 days. From the day we got there until the day we left, it was 7 days.

Almost one-third of the interviewees suggested that preparing for the worst meant having responders who were properly trained. They also considered it critical to have appropriate preparations in place at the mine site.

After questions about their initial involvement in mine rescue and their involvement in an event that did not go so well, researchers asked the respondents about an incident that was handled particularly well. One informant cautioned, "None of them start out as models. In the emergency itself, in the initial stages, there's always a lot of confusion." However, with well-trained responders and appropriate preparation, confusion can be minimized.

> SPEAKER: Our subdistrict received the report, and there was immediate response from management going to the site and beginning to set up the emergency situation, and I was called in as a part of the team that was to organize and to carry out the surface responsibilities of the functions of the emergency team. And I served in three or four different capacities. I served at mine fans where we were taking gas samples from the ventilation [and] reporting them in to the control center. I served as a recorder [for] the call-ins from these sampling sites to have the information available for the senior official to make decisions on whether or not the situation was worsening or remaining stable or improving. And I served as a guard at the mine entrance, checking people in and out that was going underground, making sure that we did not allow people who were not officially [cleared] to go underground. We had a checklist that we went by. There was supposed to be a list of those who were approved by the control center to go underground. It was to come with the group. And, in some cases, there was a failure to get the list. And we'd be told in advance that No. 2 team was going down from Jim Walters. But there'd be one or two that would be going with them. And we could not allow them to go down without clearance, and this caused some confusion. We had to get, you know, some information from control center before we could allow them to go down, because someone coming up and saying they were told to go down, you just couldn't accept. But there was a failure at times to have the proper list of people that was cleared to go underground, and this caused a little confusion. But other than that, everything went basically smooth.
>
> A system that we set up, which is outlined in the "Fundamentals of Mine Emergency Training," works real smooth. And we were going by that procedure, and this is what really set up the thought that we needed to get this type of thing in writing. And this led to the book, "Fundamentals of Mine Emergency Training," that was written as a guideline for our subdistrict area. And this is being used quite often by people in other areas as well.

SPEAKER: I guess probably the most successful ones that I was involved in were the mine fires and explosions at those shaft mines in Virginia. I mean, from the standpoint that we didn't lose any people, and even though there was tremendous damage to the mine, we didn't have a single person hurt in any of those instances. The Beatrice Mine [had] a fire—it was sealed. The mine was successfully recovered. VP No. 3 was sealed either two or three different times from mine fires; they were successfully recovered. The VP No. 5 Mine, a fire started in it, and I've forgotten how many explosions we had during that period of time before we could finally get it sealed, and then after we got it sealed, of course, it was a matter of waiting till the atmosphere was right, and then the mine was recovered, once again, with no fatalities. You know, I think you have to say that's successful.

I think a lot of it was due to the planning of the coal company. I'll have to say that honestly, because at the shaft mines of Island Creek or Virginia Pocahontas Company in Grundy, they already had formal plans for all of the materials that would be required, dimensions, everything to seal their shafts, and once something happened, it was just a matter of implementing those plans and getting them sealed. I was fortunate enough to be working with a group of damn good people. I mean, people that were knowledgeable and who had some good common horse sense, and we were immediately able to set up a good organization at the mine site and, of course, that helped tremendously. And then another thing that is of utmost importance in any of these is the assistance of Tech Support[3] from Pittsburgh with their sampling equipment and all of the techniques that they have to assist you. Without them, it would be a losing situation.

In sum, the most common lesson that experts reported having learned dealt with aspects of preparedness. One veteran suggested to responders: "Get a good procedure. Work on it. Everybody agree on it and write it up and practice, practice, practice." Another said, "It's just a matter to me of organization, and if you have the right organization, you don't have that many problems with it."

There was particular concern expressed about the need to develop a strategy for having appropriate personnel available when required. To accomplish this, staff needs must be determined.

SPEAKER: [I learned that] I would staff different. And as an example, this last accident that we had, we sent individuals up to handle the mine site, but we didn't think of our own needs within that office. As an example, we had a secretary there, this secretary, we should have sent backup for her. She ended up being the secretary, the phone answerer, the coffee pot girl, frankly, girl Friday. And we didn't worry one iota about wearing her out, the hours that she worked.

---

[3]Mine Safety and Health Administration (MSHA).

[It] didn't enter our minds, and I said if we ever did anything again, that has demands like that, I would staff from the, not just the top people, but you got to staff down below. You got to prepare for that, too.

Once staffing needs have been determined, it is critical that responders know what is expected of them before they arrive on a scene. As one expert said, "I think everybody needs to understand exactly what you're trying to do, whatever it may be." Preplanning, for both personnel and other resources, is a means of saving time.

> SPEAKER: One of the biggest lessons I learned is once you've arrived on the site, get your backup, get your support, and call for help. You are not invincible. You need help. And get your resources, and get your backup behind you...You will need them. All your resource and material, not necessarily to have the equipment on the property—underground. I mean, you can overload with a whole bunch of equipment you may never need. But you should have your resources, your checklist. If you want to call [to determine] where this equipment is and how soon can I get it here? Do I need it? That first fire that I was involved in, and I said that I was going to be the big hero, and I was going to be there. And I was there until 12:00 at night before I called for help. And then, when I called for help, it was a mad scramble, and I should have been calling for that help at 8 o'clock in the morning. When I arrived on the property, I should have made those phone calls.

As the above narrative suggests, time saved by forethought and preparation can become valuable during a response.

Regardless of how well prepared a person might be, there are always moments in an emergency when one encounters the unexpected. The informants were asked if they had seen anything during a response or had anything happen to them for which they were not prepared.

> SPEAKER: We Bureau people [operated] on the theory that a majority of the methane in the place was burned out by the explosion, and I think that's another thing Scotia taught us—that doesn't necessarily happen. There's too many variables to even consider that I guess, but I think that was an assumption made by an awful lot of people. That depending upon the weight of liberation of that mine, you have a considerable amount of time to do some things before you had an explosive mixture reoccur. Well, that's not true. I guess that was a surprise.

SPEAKER: Probably the one that hits me the most is the explosion we had at Mathies. That was the first time that something happened to a mine rescue team when we thought we had everything watched. And we thought we knew what we were doing. At that point, it was like everybody wanted to seal the mine up. I mean, their opinion was, "Lets seal it right now." And what went through my mind was, "Hey, everybody's okay. Yes, they're burnt. Something happened. Should we act drastically, or should we sit back and think about this for a little bit?" I mean, "Is there another way of going back in? Is there something that we can do?"

If I can, I prefer working with safety, instead of making snap decisions. When that happened, everybody was starting to say, "Let's seal it up, let's seal it up." And my comment was that, "I just don't know what to do." At this point, I don't know what to do, and I think we just need to take about an hour and think about this and see if there's any other things we can do. And that's what we ended up doing. And we did come up with a plan. And at first, man, everybody was getting gun-shy. And it's like, "Don't we face this everyday?" I mean, "Yes." This is the first time it happened. It scared the hell out of me, because it's the first time I experienced mine rescue team members in a flare-up or an explosion, and it was something that I never want to see again. But every day that we go ahead, every time we send mine rescue team members in a mine fire or an explosion, we can't control every location. We can't tell them that it's 100% safe, that we know where the gases are at all times. We can tell them where we are evaluating them, and we feel it's the most accurate locations, you know, and we feel that this is where the fire is at. But there's no guarantees to all that, and that happened to be a bad experience.

I sit back and look at it now, and it's like, it was a bad experience because, number one: they didn't have an updated map at the mine. We were looking at something that was so outdated. The persons that was there dealing with it from management, maybe they knew what they were doing. We were looking at what they were doing on that map that wasn't properly marked up. If we would have knew that all these situations or conditions existed, we would have probably said, "Oh no, don't do that." But they created a monster when they changed the ventilation. They did things that they didn't tell us that they were gonna do. So it could have been prevented. I hope it could have been prevented. If everything would have been done and I go back to the cooperation, not that they were uncooperative at Mathies, they were cooperative. I mean, there was a good working relationship between all the parties down there, but there was a breakdown because of inadequate maps. And the fact that nobody pointed out to us that these maps were inadequate, and the only people that were dealing with this issue was management. And we thought they were right.

Of course, mine rescue team members did some things they weren't supposed to do, too. But it's only, they didn't tell us. We didn't realize that much air was going through there. We didn't realize that there was only one entry for the air to go through, because it looked like it was all open, and air could have went in numerous directions. So it didn't look that critical. But when you narrow it down, this is the only way it could go. And right here, there was another door going back here into the shop. It's pretty obvious that you got to force all this air somewhere, and it's gonna go through that shop, and it's probably gonna take gases with it. It's gonna build up in that shop. And eventually, the fire's gonna get there, and you've got an explosion. This is hindsight, Monday morning quarterback.

SPEAKER: We had to locate five bodies, and during this time we were communicating with the surface and basically taking our instructions from the surface as we communicated. The decision was made to bring those five bodies out. We didn't have body bags—we put them in the back of a trailer and covered them with brattice cloth. When we got to the surface, I was really surprised 'cause it was dark by that time, and there was this floodlight glaring right in my eyes. You can imagine me being in this dark area with a mining lamp and then coming outside with my eyes dilated and then this floodlight glaring in my eyes. I couldn't quite understand what was taking place, but very shortly I realized that it was the news media with the cameras and the floodlight, that just as soon as we came out the door they had the cameras on us and immediately began to ask questions. By that time, the State police were there and they were very helpful. The Governor arrived, and the family members were there, too. There was some bonfires, and family members were there. I don't really understand why, but this surprised me. I guess I just really wasn't expecting this. Family members were there, obviously very apprehensive, wanting to know and really afraid to know. Anyway, I was not on the surface very long. I went back into the mine. That was my job, but I was anxious to get back into the mine. I wanted to get back in the mine—it was really easier there.

The above excerpts suggest something about coupling and communication. "Coupling," a term used by Perrow [1984], refers to the complex interactions that develop in an organization facing crisis. As pressures increase, people often neglect peripheral communications in order to focus on core task performance. The resulting increased responsiveness also increases organizational coupling. As pressure increases further, people in this more tightly coupled organizational mode start to ignore communication that is central to task performance. As key information is lost as a result, interactions become misunderstood, forgotten, or ignored. Thus, the absence of critical knowledge, such as ventilation arrangements or who has gotten onto mine property, escalates the complexity of interactions. Weick [1995] asserts that the culprit is just as much human limitation as the technological complexity on which Perrow focused.

Interestingly, Weick, though not an oral historian, argued that good stories are the key to minimizing pressure and improving sense-making during emergencies. In other words, if responders have learned through others' narratives some of the things they might encounter during an event, they will be better prepared. A few of the respondents in the present study also voiced such a sentiment. One individual proposed that mine operators have mine emergency response veterans come to their mine and talk about their past experiences, especially those regarding critical phases of events. He suggested that mine operators could videotape the presentations and then show them to personnel who would be responding to emergencies. Another expert suggested that a text-based document, providing recommendations for handling emergencies, be compiled and distributed to mine operators. This veteran believed that such a manuscript might be developed as a "how-to" manual and taken to the finest detail to ensure that all possible elements are covered. He believed that such a document would be helpful, especially if written for use by any size or type of mine. Three response experts suggested that narratives in the form of paper-and-pencil simulations of actual mine emergencies would be useful as training aids. They suggested that these simulations could be conducted in a classroom or in the offices at a mine site and used to train both command center personnel and other individuals who might be involved in the response. One expert talked about the efficacy of using such simulations of actual mine emergencies to train personnel at his company: "All the little things that seem to be taken for granted—you run personnel through a couple of exercises and you suddenly find out, 'I never thought of that.' And [the exercise] brings all these little things to the front."

References

Pearce J [1983]. Participation in voluntary associations: how membership in a formal organization changes the rewards of participation. In: Smith D, Van Til J, eds. International perspectives on voluntary action research. Washington, DC: University Press of America.

Perrow C [1984]. Normal accidents. New York: Basic Books.

Smith DH [1994]. Determinants of voluntary association participation and volunteering: a literature review. Nonprofit & Voluntary Sector Q *23*(3):243-263.

Smith DH, Macaulay J, eds. [1980]. Participation in political and social activities. San Francisco, CA: Jossey-Bass.

Weick K [1995]. Sensemaking in organizations. Thousand Oaks, CA: Sage Publications.

CHAPTER 3.—THE DECISIONS MADE DURING A MINE EMERGENCY RESPONSE

One set of questions on the interview guide covered decision-making. Regardless of how well prepared one might be, there will likely come a time when situational pressure makes decision-making increasingly difficult. Each respondent was first asked about the most difficult situation he had ever found himself in (or decision he had to make) during his work with mine emergency response.

SPEAKER: That one is very simple. At the same time we were trying to recover a roof fall fatality victim at Cherokee Mining Company, we had a victim trapped by a roof fall on a shuttle car at Canada Coal, and I had to split the mine rescue team, and I took half of them and went to Canada. We left Cherokee, and I left half the men trying to recover that one. We knew we had a live man trapped. We worked about 5 hours trying to get him out, and the roof was so unstable that we experienced three more roof falls on the shuttle car while we were attempting to get the victim out. At one time, I made a decision that I'm glad that I didn't have to do. The only way that I thought we could get the man out was to take a torch and cut the deck off the shuttle car and let it set down. Hindsight, 20-20, I see now that I would probably have killed the man. And I asked my team, I said, "Let's try one more time" to get him out, and everything from that minute fell in place. We went straight down through the rock, and within a half-hour—after 4½ hours of fighting it—within a half-hour we had him out. But that was one of the toughest decisions, and I sit back and think now, had I gone ahead and followed the initial decision I made to try to cut the end of the car out with a cutting torch, I would have killed a man. There's no doubt in my mind, and we were lucky. We managed to get him out, and he's alive and well today.

SPEAKER: I think the one that was the most emotionally charged was the one in 1984. We went down, and this poor chap was pinned. It took us almost 24 hours to dig him out. And knowing full well that when you dug him out, you're not going to save him anyway 'cause he was crushed to the point that all that was keeping him alive was the fact that he was crushed in. And as soon as you release that pressure, once the body tries to come back to normal, unless you got those pressure trousers ready to put on him, you're going to lose him. And we can all still hear that poor boy, you know. He never screamed. He never yelled. Never, never screamed, but just [pleaded], "Take me out." I was the most emotional. It's hard. I get emotional now just thinking about it. You know, there were a lot tears shed that day. I'm almost ready to cry. But again, that's what it's all about, isn't it? You're going around to rescue someone.

SPEAKER: When I went on surface, what happened is the silo collapsed on an employee and he was injured fatally. It was the first time this ever happened, and it took a little while to get organized. The delay wouldn't have cost the man's life because he was dead on the time, so he did not suffer. But it did take a while because we weren't used to having emergencies on surface; underground only. And all of a sudden, this was on surface. Do we use the fire crew? Do we use the mine rescue? By the time decisions were made, you're talking about several hours. Not blaming the company, not blaming individuals, but it took that experience to learn by it.

We know that we can't reach him. All we know, we've seen him. Well, we finally got a hold on him—we know he's dead. But we can't get him out, because he's crushed. And now we've got to make decisions to use torches or different types of equipment. And you still don't want to, you know. You've got to make a decision. You can get him out real fast if you cut him apart. He's dead anyway. He's been dead for 3 days. But I know I didn't want the body unintact. Therefore, it took us a good 12 hours to get the person out without damaging that person itself, using different types of emergency equipment to get him out.

SPEAKER: The situation that I had to face is when I had to call a guy's family. I can think of that today, and it still bothers me. You pretty well know that this guy is dead, but you don't want to say that, and you can't say that, if you don't know 100% positive. And then it's not my—I don't think it's my—I shouldn't be the one saying that anyway. Not at that time. Before I'd say that to the family, I'd have to sit down and think for a few minutes to see what words I'm going to put that in. So that was the hardest thing that I done during the whole [response] was calling the family and letting them know that this guy's not going to—he's trapped. Well, he's not trapped—he's unaccounted for. I tried to use words that had the same meaning.

SPEAKER: The hardest decision that I ever made, and I had to do it instantly, was withdrawing the miners from Wilberg[1] when the fire had gotten out of control. The way that stuation was set up, we were actually fighting the fire from the inby side of it. We were traveling up

---

[1] Wilberg Mine, Emery Mining Corp., Orangeville, UT, December 19, 1984; 27 killed.

by the fire, circling, and coming on the back side of the fire to fight it. And the fire was breaking out in several places in our area. Our reports from the fresh air base was, "CO's going up in the fresh air base. We got 40 some parts per million in this fresh air base." I'm going from memory; I may have it a little bit out of line. Then they said, "People are panicking up here. What do you want us to do?" Coal miners and support people who were in there actually left that area running towards the outside in a panic, and we're talking about 50 or 60 people was in there at that time. We had several bodies in the fresh air base at that time that we had brought down, and the people that we had in there did not take time to pick them up and bring them out because they didn't feel like they had time to do that. I mean, they had actually left there in a panic. As a matter of fact, when they left there they were running towards the outside of the mines, and they were underground, oh, a little over a mile. I gave instructions to bring all my people to the surface, and the company opposed it on the spot. The company said, "If you bring them out, we'll lose the mines." I said, "Well, you're going to lose them all, if you leave them there."

The vice president of the company met them at the portal, trying to get them to go back into the coal mines, and our people wouldn't go till I met with them, and we gave them instructions that we would go up to the fire and fight the fire from the outby side like we should've been all along, and we'd furnish them with anybody that they needed to do that—support, but we would not allow our people to go inby the fire, because they were not mine rescue team members, and the mine rescue team members would have to make their own decisions—we don't make decisions for mine rescue teams, but our people was not trained in [the use of] apparatus, and we wanted them out in the safe area.

SPEAKER: I would say the hardest decision was the sealing of the fire area at the Orient No. 5 fire when we knew there were three people unaccounted for. This was a tough decision. It was tough for the families. It was tough to come out and face the families and say, "We've sealed the mine down there." And they had three people—very frustrating to the families and very frustrating to the people there at the mine, especially when it's people that you've worked with.

SPEAKER: To seal a mine—with people in it. Yeah, Scotia. After the second explosion, we were able to get in through a shaft, and the bodies were observed, but we didn't know what our available ventilation controls in the mine were like. We didn't know what had caused the second explosion. There were just too many unknowns at the time, and that was the reason we decided to go ahead and seal it and stabilize everything. I guess I was probably the strongest for it.

But, we just felt like we were doing the right thing and the safe thing. I think the first day, we spent trying to see what we could find out underground, and in the second day, we made this decision to seal and began sealing.

SPEAKER: The most difficult one was at Fire Creek, when we sealed the mine and left people in there. The reason that was more difficult than Wilberg is because we had a fire raging at Wilberg, and there was a lot of people making the decisions, you know. There was only three people making the decisions at Fire Creek, and I was one of them. And even though it was short-term, about 48 hours until we got a hole drilled down and could get back in there, that was the toughest one to make. But it didn't take me long to make it. I didn't really hesitate. I thought about it maybe. I thought about it before we ever made it. And within 5 minutes after we started talking about it, I was in agreement to do it. Because, one thing, I thought we could get back. It was easy in one way 'cause I thought, "Well, we could get back in there pretty quick if we need to 'cause they're not very far back, and it's not a massive mine like we had at Wilberg."

SPEAKER: I was notified, and by the time I got over there it was about midnight, and they had just gotten everybody out, and of course the discussion was going on. You know, "What do we do now? What do we do now?" And the safety director said, "Well, what I would like to do is wait until the morning and go in and take another look at it." My response was, "With the amount of methane that's in this mine, with the fact that you have a fire out of control, there is no way that you're going to go in there in the morning and take a look at it." The president of the company and the safety director came about that time and we discussed it, and they said, "Well, what are you saying?" I'm saying, "Well, you seal the mine. There's no way out of it." Now, this is not a decision that's made concerning life, but it is a decision that's made concerning economics. I mean, here is a mine employing several hundred people and you're saying, "Seal it—period." You can't make decisions like that lightly, and even though they don't involve life, there's a lot of stress and strain in making decisions like that.

This set of accounts suggests that situations concerning people (either alive or dead) and livelihoods are seen as the hardest conditions under which to make decisions: "There are tough decisions made concerning people, but there are also tough decisions made concerning other things in a situation such as that." As the literature suggests [Weick 1995], not only do complex tasks start to become senseless as arousal increases, people tend to abandon newly learned behaviors (such as those acquired in recent training) and fall back on older responses: "When something like this happens, and you get there, you're not a damn bit smarter than you were the day before. You don't know a bit more than you did the day before, and the only thing that you can do is use some good old common horse sense." Other factors that affect the decision-making aspect of responder involvement will be explored in the remainder of this chapter.

Several of the interviewees thought that the focus should be on interactions between responsible individuals in the command center. When asked how decisions are usually made and who makes them, the veterans generally responded as follows:

> SPEAKER: Central. In other words, there will be the mine manager, the one in charge. But he doesn't make no decisions without consultation. If it's ventilation, it's the ventilation technician. If it has to do with blasting, then the blasting foreman comes in. If it's mine rescue, then it's myself. He's never made a decision without consulting the people well trained in that situation. So it's very good. There's not just one man making completely a decision, which, you know, could not be, it wouldn't be fair for that individual to make all major decisions. He's doing it in consultation with the people that are trained.

> SPEAKER: I am in charge until the arrival of the Commissioner of the Department of Mines, according to our State law. I solicit ideas. I am nowhere near the smartest guy there. I realize that. There's always MSHA personnel on-site. They have some very qualified people [including] the one I told you about. Hargus and I've worked closely on several recovery operations. I have great respect for his opinion. I solicit ideas from even my mine rescue team members. Most of the time, I will incorporate an idea that somebody else comes up with.

> SPEAKER: Well, every one I worked at, you always have somebody in charge of each group. There's somebody in charge of the union's people. There's somebody in charge of the company. Somebody's in charge of the State and Federal. And the best way that's handled is, it's the company's coal mines, and if you got the people there that can make those decisions, it's better to allow them to draw up the plans. And sometimes we even meet and discuss options. You know, "What do we want to do next," and we all sit down and discuss what needs to be accomplished, and usually a good company who has got the personnel to do it, and expertise, will say, "Okay, I've got a pretty good idea on what I think most of the people in this room wants to do and how," and they'll go out and adopt a plan and put it in writing. Now, that's the plan of what you want to accomplish and how we're going to do it in broad outline, but usually when you throw that into place,

you've got to change and roll with what you're finding. Be able to adapt to it, you know. I found out that what really makes it hard is if the company tries to get all the details and do it exactly the way they say, to start with. If you just say, "We're going to start a team exploring—the fresh air base is going to be here, and we'll act accordingly," it expedites things.

SPEAKER: At that particular event, you had a group of people who were a district manager and a couple of subdistrict managers, and there was a representative from the company, the operator, that ended up coming together in a group. And they were the more senior, experienced officials, and they would congregate and talk about what it was that they were facing. They would bring [MSHA] Tech Support in, and Tech Support would provide them with advice on their analysis of data gathered from fans, from air bottles, from whatever it was that they were particularly doing. And they, frankly, just got together around the table and they just talked, and generally one person would come to the front and be in charge. I had a neighbor who says in the absence of leadership, leadership will appear. And he's right, that no matter, you don't have to have any title or anything, but some strong person, some forceful person will come in and he will be recognized as this, and he will be the leader. You can't do it by title, they don't care if you're a district manager or whatever it is, but this person will come forward, and there it happened to be John Weekly. And the others fell in line behind him as a team that became advisors, because he was pretty much in control. He was just a person that was recognized as in charge. He didn't wear a hat, you know, there were senior people to him at the mine site that were district managers and stuff, but he was just recognized as that. For whatever reason, some person will come, will be forwarded, and he may not step forward, it may be that others step back and leave him in a leadership position, but that always happens.

The above comments on how decisions are generally made during an emergency response coincide remarkably with the characteristics of effective leadership derived by Kowalski from studies of mine fire escape behavior conducted at the Pittsburgh Research Laboratory (PRL) [Kowalski et al. 1994; Vaught et al. 2000]. The leader in each case was described as an individual alert to his environment, attentive, and discerning. It was also thought that this person might excel at incidental learning, retaining information that was instrumental to the escapes. A second characteristic of each leader was the manner in which he took charge. The emerging leader did not "muscle in"; his leadership developed in a natural way. Third, the leaders were decisive, yet flexible. They made decisions, yet if circumstances changed, they adapted. Fourth, the leaders were open to input from others. Fifth, there was a logic to the leadership. Decisions were appropriate and congruent with the available information.

Decisive leadership and appropriate decision-making become crucial at certain moments during a response. It was hypothesized that these instances might be common to most emergency events. The veterans were therefore asked: "Are there certain critical points common to many mine responses, times when certain actions should be done or decisions made?" They were further requested to discuss some of the most important decisions that have to be made.

> SPEAKER: I'm not sure that there are times; there's probably timeframes, you know, that after a while a decision has to be made. My response to that would be: yes, there are times that decisions have to be made that would be within a particular timeframe, that you realize that not making a decision is really a decision not to act, and the question is, "Do we need to act, do we need to act a particular way?"

> SPEAKER: I was talking about a window of opportunity. And when these do occur, you've got to grab the opportunity and to do what it is you have to do. You have to be constantly watching conditions to make sure that your people aren't being put in harm's way. So the people involved who have responsibility for making decisions, this is the most critical thing they have to do. And it's one of the things that we have so much trouble teaching people, is to watch trends, not numbers. Everybody gets excited by a number, and it's so hard to get them to understand that very low numbers can be extremely dangerous if they're increasing. I told you about the pseudofire we had at Jim Walters. Almost a year later, close to the same area, we had a real fire on a longwall. And you'd hear the pops in the gob as pockets of gas would get ignited as we were sealing. But we had a rule that we use our Zabetakis nose-curve as our guide. And when we had three consecutive rises, no matter where they were, we pulled back. And it's interesting to see hairs grow gray when you've had two consecutive rises. And everybody's waiting for that third, because when you take three rises at 20-minute intervals, [you] should intervene. There's certain things that you can do. You're buying time. And people, if you're following what the trends mean, you can make the proper remedial effort. And so, it's so necessary to teach people trending. Yeah, these are decision-making devices.

> SPEAKER: Levels of gases get to a critical stage—a decision has to be made. Do we pull 'em out of the mine? Do we pull 'em back to a certain location? When changes occur, like ventilation changes, what protections are we going to follow? How are we gonna do it? Where

are we going to bring them back to? So I guess anytime there's a change in what's happening, it's very touchy. And you have to make the right decision and get 'em out. Of course, when you have the pops, and it happens to be that everybody is still there, then you start wondering, because the samples didn't come until a half an hour later that showed what happened a half an hour ago. It was too late, and that has happened. Maybe it was a lack of manpower or lack of whatever, but it seemed like somebody in the computer room that they had set up right at the mine may have been getting this information on a regular, quick basis, but everybody wasn't. And that's one of the most important things that I think that we have to deal with as persons that are in charge of an operation, is to make sure that those persons that are underground are safe. And the best way to make sure they're safe is to evaluate those gases in the best possible location.

SPEAKER: Well, I guess when certain things occur, you got to make a decision, you know. If you have a team that's working and something is discovered, then you got to make a decision in order to go on, or to withdraw. Now, some of the times they don't do it according to the book. In other words, the rescue teams have got certain rules, but you don't go by those things. You do to a degree, but there's a lot of times you do things that I don't guess you want to write home about. You know what I'm saying.

SPEAKER: I guess one of the cardinal rules is: you don't do anything about ventilation until you know where your men are. You don't shut it down, you don't disrupt it—you just leave it alone. Do nothing, and then go down and find out where your men are, what the situation is. Another mine rescue rule is that you never split a team. You're always traveling as a five-man team. That one seems to be a kind of a touchy one, because the rule is, a fire call comes to the mine, nobody moves until you've got two teams on the property. You got a team for underground and a team for standing by. And sometimes it takes a fair amount of time to round up 10 men. And we've always said to ourselves, "Well, what happens if there's somebody's life at stake, and you've only got three men?" And the answer to that one is: the circumstances will dictate what you have to do. And if this guy is a couple hundred feet, and you know he's a couple of hundred feet away from the shaft station, well, you're going to go get him. If he's a couple thousand feet away, and you are going to jeopardize those three men to go and get him, then that's a cold, hard decision that we have to make. You've probably lost one man, and let's keep it at one man. We're not jeopardizing any more with the remote chance you might save him. Those are some cold, hard rules of mine rescue, because I've always said, "You can call these men from home; they were safe at home. You call them out, and you send them down to fight that fire, and they better come back to you." If we're going to lose anybody, unfortunately, and this is a very hard thing to say, if we're going to lose anybody, it's going to be the poor man that's down there, not the mine rescue man going to get him. But those are the kind of rules that we pretty well stick to.

SPEAKER: I have never seen two the same—there's no way you can have a rule of thumb on any recovery operation. Every one I've ever been on, even if it's roof falls, each one is different. You have to look at the situation you have, decide the best approach. Sometimes you don't even know where the body's located.

SPEAKER: I think each and every emergency situation stands on its own merits, and the decision-making [at] that particular one may not be compatible with another one. You'll find that they're practically all different. It's like a head-on collision out here on the highway. No two are going to be the same, even though they're head-on collisions.
There's some pattern, but I think every mine fire you have, every explosion you have, has to be handled as a different situation. I don't think that you could set a pattern and say if there's a mine fire or explosion here, here's what I'm going to do.

SPEAKER: Every one of them is totally different, but still in all, there is a pattern that you go by. But, statistics is what guides your decision-making. It's reaction to an action. That is the best way that I can tell you about it. If the gas is building in certain places and you can control them, then control them. If you can't, then you have to make a decision to go at it a different way. This is generally decided by whoever's leading, but I can assure you that he don't make it on his own. I always have a backup group that's sitting there telling me what's going to happen next. They're predicting from a set of facts— air readings, gas testings, procedures—there's something going on that they're guiding me. It's up to them, when they see an action, to make sure I react to it. The only way I could react to it is by knowing what's going on, and that's the way I have handled [any] that I've ever handled, and I have handled them for years.

The individuals responding to this question expressed themselves much as one would in grappling with the differences between art, which is a skill obtained by the study and practice of certain principles, and science, which involves activity based on methodology and discipline. *Each emergency is unique, but there is a pattern to go by. There isn't a "time" to act, but there* are timeframes within which particular actions have to be taken. One must follow the rules, but be alert for windows of opportunity. The "book" is sacred, but we don't always go by the book. As in most realms of activity where art and science are intermingled, there must be some means of orchestrating the two.

Cognitive psychologists, in trying to understand the way in which expertise is acquired and used, developed two related theories that explain how this orchestration might work. The first theory of expertise is called "chunking theory." Chase and Simon [1973] defined a chunk as information that has been grouped in some meaningful way in the subject's long-term memory so that it can be remembered as a single unit. Therefore, a multichunk unit will take up no more room in a person's short-term memory than a single unit of unchunked information. Because the short-term memory is only capable of storing a half-dozen or so units, space is limited. It is thus to one's benefit to chunk. The difference between a novice and an expert, then, resides in the fact that the latter has acquired a large database of chunks so that he or she does not have to respond to a situation using one single unit at a time, and therefore has superior problem-solving abilities. Still, some psychologists believed that these notions did not explain how an expert can solve problems in chaotic surroundings. To address this aspect of expertise, Gobet and Simon [1996] introduced the notion of "templates," which are retrieval structures with an unchanging core of chunks and a set of slots that can be altered rapidly as they are filled in by context or by additional information from the environment. New information is processed according to how it fits into these slots. The resulting schema can be used not only to interpret a situation, but also to predict what is likely to occur in the environment.

Template theory explains how an experienced person can arrive at the scene of an emergency, take a quick reading, and begin problem solving. He or she already has part of the picture stored in templates, so it is just a matter of picking a template that best fits the situation and filling in the slots with information specific to the problem at hand. Klein [1989] noted that experts at an emergency do not consider a broad array of possibilities, but rather focus on what "type" of situation they are facing based on past cases (in other words, their templates). They are thus able to embark on decision-making strategies that allow them to cope with a host of ill-defined contingencies in a timeframe that would not allow for deliberate and judicious steps to discover and implement the "best" choices [Perdue et al. 1995]. In essence, the expert, in choosing a template, selects one that is "good enough." The template is then used to guide decisions that are "good enough" to work in the circumstances that he or she presently faces.

Whether guided by templates or some other cognitive schema, decisions are not made in a vacuum. The veterans were asked how decisions are made during an emergency response, and who makes them.

SPEAKER: Well, we touched on that. Usually, it's a unified situation with State, Federal, union, or representative of the miners, whoever it might be, and management. But again, the bottom line is management. They're responsible for everybody and everything. Even though there might be Federal or State people, they're still responsible for them.

SPEAKER: Ideally, you would like the State, the representative of the miners, or the union representative, whichever it might be, and the company, and an MSHA official. Now, it doesn't always work out that way. Generally speaking from my experience, the people who make the decisions are generally the MSHA people and the State people, and, all too often, it winds up with MSHA making the decision. I remember several years ago when we had sealed the Beatrice Mine. What I'm going to say doesn't refer to the Virginia Department of Mines and Minerals now, because the gentleman that they have there as chief inspector is a very cooperative person, and he and MSHA work beautifully together. But this was prior to his time. I remember that we were to have a meeting to discuss reopening the mine, and I had Jim down from [MSHA] Tech Support in Pittsburgh. We had gone over all of the samples there, how everything looked, and made the decision that now just was not the time to open the mine. I remember calling the Chief of the Virginia Department of Mines and discussing it with him, and he said, "I'm right with you. That's exactly the way we will do it." We went over there the next day, and I saw him before the meeting started, and his attitude was exactly the same way. "That's it; we just won't open it now." Well, we get in the meeting, and the company goes over all of their thoughts and theories on it: why they think that it should be open and gave their conclusions as to the decisions that they had arrived at, how they had arrived at them, and so forth, and they asked the Chief of the Department of Mines what he thought about it, and he said, "Well, if that's what you want to do, that suits me fine." He looked at me and said, "What do you think?" I said, "No way are you going to open this mine now." So, I mean, those are some of the situations that you get into. I've had that happen to me several times. And those are tough, those are tough.

Sometimes, according to the respondents, decision-making goes fairly smoothly; in other cases, it doesn't.

Decisions are made more easily when they are informed by moments that have gone before. During their interviews with PRL researchers, the experts argued this topic area should include not only training in the development of an emergency response plan, but the training of personnel in its implementation as well. Implicit in their observations was the notion that it is critical to arrive at a well-designed document that has been put to the test.

SPEAKER: Greenwich Mine—they had a plan in place. They had an explosion. They started neatly, called the rescue people, had them on scene. They called ambulances in, paramedics, and I support that. Alerted people for supplies that was going to be needed, especially a large amount of supplies. Again, it depends where the explosion's at, how far it's in the mine. But generally, when you have an explosion, it takes out stoppings, and you need canvas, you need lumber to put this, the canvas checks up, you need the foam spray that you spray on it to keep it from leaking. 'Cause when you put these things up around the ribs, or sides, they leak a good bit, and you take this foam and spray it around so that you can keep your air, when you get up there a certain distance with all this leakage, you finally end up with no air. I think that what they did at these mines, they had places where rescue people were assigned to stay so they could get their rest. The worst thing is to keep rescue people in the lamp house where people's walking around, people are hammering, people's talking, making noise where the rescue teams can't get their rest, where you have the press running around, where you have families running around. I think that a person should be assigned and say, "I will take the responsibility to see where the rescue team's going to stay at. We will have a place for the families to stay." Generally, you have the Salvation Army or the Red Cross come in, and you want some place for them to stay, and you try to get them away from where the action's at, 'cause if you start bringing a large group there, everybody's running around and they just interfere with one another. If you can, [move] trailers in, or use the warehouse or supply house, or something to get them away from right where the shaft's at, and the main office where people are calling the shots. Greenwich—they had the State police there to fence off the area. I think that at Greenwich, it was. I didn't totally agree with it. They didn't let anyone in, but they didn't talk to the press, and I think that they should have had a company person go down and talk to them. I think that you've got to make arrangements to have trucking companies available to pick stuff up that's flew in to an airport, so you can move the material immediately. Again, it depends on the number of bodies, but if you have a large number of bodies that you're bringing out of the mine, I think they should be moved from the mine immediately, not even put in a room, that there should be an ambulance there, and put in the ambulance and taken to wherever they're going to perform the autopsy, or someplace. When you keep them at the mine site, it's a problem.

In addition, the importance of practice was reiterated as a basic theme. The veterans suggested that some valuable experience could be gotten during activities such as mock emergencies, although at least one felt compelled to differentiate between experience and proper training.

SPEAKER: I'm a great believer in plans if they're flexible enough, good enough. But what good's the plan if we don't have the training associated with it? We had [a] mine explosion down here in southern West Virginia. And you have one of the best companies in the United States there. You've got MSHA down there, and no one went into the bleeders, to monitor the bleeders. They claimed, "Well, we were monitoring at this point." But you look at the ventilation, and I'm not sure what you're monitoring at that point. But we allow them to change the ventilation without having to check the bleeder. This fire, the explosion, and the flames involved the gob. It's like looking at a coin on one side. If it's a false coin, it's got heads on both sides, or tails on both sides. And that's this day; we're talking about 1992. This company has wonderful plans. I happen to know that they've been involved in these mine emergency training programs where they have this mock stuff going on. The time came, they make ventilation changes—a basic no-no. Scotia's another good example of people doing things that shouldn't be done. Very sad—I cry about that. At the same [time] we were going into court on Sunshine, and I was in Boise, Idaho, testifying in court, we had our second explosion. And I cursed, because none of us who should have been there were there. So we feel very guilty about that today. But you see this going on constantly, and I don't expect changes. In fact, a number of us expect only the worst in the future.

With regard to the last statement about expecting the worst, experts were asked about issues that should be passed on to future responders. It is interesting to note that, during this discussion as well, mine emergency veterans stressed the significance of preparedness with respect to (1) having a well-designed emergency response plan and (2) practicing for emergencies.

References

Chase W, Simon H [1973]. Perception in chess. Cognit Psychol 4:55-81.

Gobet F, Simon H [1996]. Templates in chess memory: a mechanism for recalling several boards. Cognit Psychol 31:1-40.

Klein G [1989]. Recognition-primed decisions. In: Rouse W, ed. Advances in man-machine systems research. Greenwich, CT: JAI Press, pp. 47-92.

Kowalski KM, Mallett LG, Brnich MJ Jr. [1994]. The emergence of leadership in a crisis: a study of group escapes from fire in underground coal mines. Pittsburgh, PA: U.S. Department of the Interior, Bureau of Mines, IC 9385. NTIS No. PB94-176435.

Perdue CW, Kowalski KM, Barrett EA [1995]. Hazard recognition in mining: a psychological perspective. Pittsburgh, PA: U.S. Department of the Interior, Bureau of Mines, IC 9422. NTIS No. PB95-220844.

Vaught C, Brnich MJ Jr., Mallett LG, Cole HP, Wiehagen WJ, Conti RS, et al. [2000]. Behavioral and organizational dimensions of underground mine fires. Pittsburgh, PA: U.S. Department of Health and Human Services, Public Health Service, Centers for Disease Control and Prevention, National Institute for Occupational Safety and Health, DHHS (NIOSH) Publication No. 2000-126, IC 9450, pp. 170-171.

Weick K [1995]. Sensemaking in organizations. Thousand Oaks, CA: Sage Publications, p. 102.

CHAPTER 4.—SOME SPECIFIC ASPECTS OF MINE EMERGENCIES THAT AFFECT RESPONSES[1]

Chapter 3 discussed moments in an emergency response during which decision-making strategies have to be used. This chapter details several specific aspects of any given response that might properly be called "background problems" or distractors. These variables, while not primarily instrumental in the response process, are nevertheless stressors that impact not only the quality of decision-making, but also how well the emergency organization functions.

The emergency response organization is not solely a technical system or a social system, but a structural integration of human activities around various technologies. This portrayal of organization casts its management as a process of maximizing the system's chances for success by facilitating internal and external relations while promoting organizational effectiveness. However else it might be characterized, an emergency scene is, for most or all of its participants, a workplace. The fact that it is such a dynamic environment has often obscured this truism from researchers.

Emergency personnel, like all workers, carry out their duties within a context composed of discrete elements. First, there is the event itself, which presents responders with a need to behave in a proactive manner. The simplest way to express this notion without delving into a discussion of dialectics is to state that such a situation constrains workers to be as much "actors" as "acted upon." Second, the emergency occurs within an existing societal structure that is composed of specific social units with rules and forms of association. Within this framework there must be, of necessity, an emergent organization functionally dedicated to the accomplishment of a relatively narrow and short-term mission. A response, then, may take place within a nonmaterial context of values, norms, beliefs, and behavioral expectations that are often in conflict with each other. Third, there is a technology that must be understood and used appropriately in order to accomplish collective goals.

It has already been suggested that demands on a person acting as one of the directors of an emergency response are many and varied. The leaders are exposed to numerous stressors such as time constraints, information overload, and, especially during the initial organizing stages, a rapidly accumulating array of subtasks. In major operations, as the study informants have pointed out, decision-making is shared so that no single individual has complete control over the current situation. Salas et al. [1996] characterized this state as a condition possessing work entailments that "evoke an appraisal process in

---

[1]Some sections of this chapter were originally authored in collaboration with Dana C. Reinke, sociologist, Pittsburgh Research Laboratory, National Institute for Occupational Safety and Health, Pittsburgh, PA.

which perceived demands exceed resources and result in undesirable physiological, emotional, cognitive and social changes." The notion of demand versus resource (as it applies to human resources) is key to emergency management. Many demands upon the available participants stem not so much from the response problem itself, but from background problems that may or may not become critical yet must nevertheless be sorted through. The more efficiently and effectively this is done, the more successful a given response is likely to be.

In any mine emergency, as Mitchell [1990] so succinctly pointed out regarding fire, "time is not your friend." Except for those individuals who happened to be on-site when an event began, all other responders would have had to first be notified and then might be required to travel significant distances to the rural area where the emergency was taking place. The informants were asked a series of questions regarding the elements of time with which they would have had to be concerned: How long after the discovery of a problem at a mine site were you usually contacted? Who contacted you? How long after you were contacted did you usually arrive on the scene? How long did you usually stay on-site?

SPEAKER: Within minutes, within hours—it's never the same. But usually, if I'm on the property, it's within minutes. If I'm off the property, then it's within 15 to 20 minutes. I'm the second one on the list that they call out for any emergency.

SPEAKER: It could be as little as 45 minutes and as long as 2½ hours, maybe, if I had to drive.

SPEAKER: At the Grundy disaster, I was contacted probably within 4 hours, and it was approximately 12 hours to 16 hours when I actually made it on-site because it was 150 miles away.

One mine fire, I was within 15 minutes of the mine site itself and we were, we were contacted. And it was an odd thing on it. We were doing chest x-rays. There was a program from the, I think the National Institute of Health about doing chest x-rays on coal miners. And we were instructed to go to this, and it so happened that this site was set up at a union hall. And they were doing union members, too. So we, would go to this to do this chest x-ray and while we're there one of the guys that we, that ends up knowing us says, "Have you been contacted yet? They've been hunting you." I said, "No, we didn't know anything about it." So we go in and get on the telephone, and we are less than 15 minutes from the mine site itself, and they said, "Well, we got the mine on fire." So I said, we go out and get in the car and there was four of us, we go get the car and drive there. So we're there within 15 minutes.

SPEAKER: Calling the mine rescue station is one of the first phone calls that come from the mine. It usually goes Mine Manager, Safety Department, Mine Rescue. We have a policy that when the team does arrive on the property, we expect them to be ready to go underground in 15 minutes. I've been called. I arrived one evening, one call in particular. I was called 12:30 at night. It's about a 5-minute drive to get to the mine site. The men arrived there at five to 1:00. Twenty-five minutes later the first team was, what we call, "ready for briefing." A couple of other incidents, the one, the latest one, which is only about a year ago, we were called out at 3 o'clock in the morning. And at 3:30, the first team was ready for briefing, but the company officials weren't ready for the team. We were ready before they were, so we're called out relatively early.

In the immediate area where I'm in—on up in that area and all of the mines in that immediate area are within 10 minutes. I can be at any of the mines in 10 minutes. Now, there are other mines that will take me 25 to 30 minutes away. I think that I can safely say in the mines that are in the immediate area right now that are operating, are within 30 minutes. In fact, I can be to any of the mines in less than 30 minutes.

SPEAKER: If I had responsibility in that geographic area, I would have known about it right away, one of the first. If it was an area that I didn't have immediate responsibility, then it would have been later. Like example, Finley, which our field office didn't cover, but they needed some help, so they called us. I learned about it as soon as I got in from the mine, but they had just received a call there in the office just a few minutes before that. So this—the explosion occurred just a few minutes after 12:00, and the reason I remember that, it sort of shocked me when I realized that I noticed the wristwatches that some of the guys had on, and I don't understand this but they had stopped, and I don't know whether it was just a tremendous amount of force that had done that, and I noticed that several of them had stopped at, like, 8 minutes after 12:00, 10 minutes after 12:00, something like that. So this would have been probably 2 hours, 2½ hours after the explosion occurred that we knew about it at Hazard. And the call had come from Joe Malesky's office here in Norton. Although the explosion was over in Kentucky, this was their district headquarters. So it would range from immediate notification to 2, 3, or 4 hours later depending on the area of responsibility.

SPEAKER: I'm usually contacted when all else fails. I'd say that's the general rule. Because you really don't need me until you can't do the job yourself. If you can do it, you don't need me. I do have a couple of companies who call me fairly early in the game, Mapco being one of them—probably now, three times—immediately. And Ziegler Coal's another that does that.

Well, I'd say anywhere east of the Mississippi, I can be there within 5 hours.

SPEAKER: We had a procedure when I worked for the company. That was a standing thing at all the mines that the head safety man and the head operations man were to be notified pronto—right away. And we usually got there as quick as we could. Naturally, if we were out of town, you know, that wouldn't work, but it was quick.

SPEAKER: We are within 1 hour of travel time of virtually every mine in our district. So notification, if it's given to the office, we can be to almost every mine in the district within an hour.

SPEAKER: We were generally contacted by the company within just a very short period of time, and I can't really give any timeframe, but I mean it was very soon after the things have happened that I've been contacted. Well, you know, that depends strictly on where it was. Over in Buckhannon County, it would take us about 2 hours. To get over there, and that's where most of our problems, of course, McClure is about an hour drive. I'd say from an hour to 2 hours from the time you were notified.

The above answers to questions about notification and arrival times point to two differences between rural mine emergencies and urban emergencies: (1) a mine emergency often unfolds over a longer period of time and (2) most resources to deal with a mine emergency, both human and material, will need to be imported over sometimes long distances. Nevertheless, as their responses show, the time from discovery of an event to notification and the time from notification to arrival on the scene were highly variable background factors for the veterans.

Another dimension of time that may be problematic at mine emergencies is what often seems to be less-than-optimal work scheduling. To get some sense of this aspect of organizational behavior, the subjects were asked how long they usually stayed at the scene once they arrived.

SPEAKER: I had gone like 28 hours and not left. But I don't think that's a good idea. I think, probably from 12 to 14 hours. That's when the change needs to take place is within that time period. And that doesn't mean I need to go down to a hotel 20 miles away. That means, I go to a back room where nobody will bother me. Just sleep 4 hours.

Then something happens. Somebody comes wakes me up in 2 hours. And I say, I'm good for another 2 hours, but that will be it. That's all I'd be good for. So usually 12- to 14-hour shift is, I think, probably anybody should be around. It's hard making all the, you know, the tough decisions. But if your things are slow, then I'm not, I may be there 3 hours, and things are going to be slow. People are taking readings. We're waiting for people to bring bottle samples back. I know nothing's going to happen for the next 3 hours. And I've only been there 2 hours. I'm going to go somewhere right then and rest. I'm going to tie myself up on the front end as well as in the back end.

SPEAKER: Well, I'm the type of guy, you can't compare that. I'm the type of guy that will stay as long as I feel comfortable. But sometimes the decision has to be made by management that I've been there long enough. It's time to go home. We try and organize, and we try and make sure that we figure in 8 hours, more than enough. But in a few circumstances, it had to extend a bit, only because of, you know, of the circumstance. But now, we're trying to work very hard. Every time we have a response, we do not, definitely do not want people more than 8 hours there. We figure the decision-making are better fresh within the 8 hours. Anybody that has to stay 24 to 30 to 40 hours, you know, there's no way that you could be able to make a [good decision], and we know that. So I don't say it never did happen. No, it did. But we try every time to [avoid] it. That's where the guy in the control center has to come and say, "Hey, you're going home." The next guy comes in.

SPEAKER: Well, I watched J. W. go into 72 hours at one time, and that's too long. He would tell you that's too long, too. It's almost emergency-dependent. If it's someone trapped, if it's a mine fire, if it's something else. We, from burnout, from doing things like that, have tried to establish that you won't stay longer than 12 hours. And if I had to recommend it, I would recommend it to 12-hour shifts. If you do 8-hour shifts, you got three sets of individuals that have to be briefed and so 12-hour shifts leaves two groups of individuals running something. I think 12 hours, if it's going to be many, many days, can really get old, too. But 12 hours we've handled, and very well, on several occasions now.

SPEAKER: Command center people are pretty much 12-hour individuals; others, depending on how harsh your task is. If you were going to go with the mine rescue team, we only work with mine rescue teams under oxygen. We try to limit that to a maximum of 2 hours in, in every 12-hour period 'cause you can really wear mine rescue teams out, because working under an apparatus is very, very physically demanding. And if you're going to go on something for several days, you can exhaust your people very quickly. So you're trying to stay aware of all of that, and you go on scheduling. MSHA people, again, I think it's dependent upon what their jobs are. We have been fairly successful with 12-hour shifts about this changing MSHA people out. What you're kind of expected to do is to come there an hour early so that you can get briefed about the person that you're replacing. So it may almost even get to be a 14-hour. But once the initial 2 or 3 days of it takes place, the transition will be a lot easier.

SPEAKER: We work on a shift cycle. If it's going to be, I arrive on the property, and I'll be there usually about 8 hours, and then my replacement will come in. Usually, I work on an 8-hour cycle. A couple of incidents, I've been there longer, but we've learned the lesson that after 8 hours, you should be bringing in fresh troops. You get tired and you're not thinking clearly. You'll make a—maybe make bad decisions, and we—I've learned that lesson very quickly. There wasn't a backfire. I was going to be the big hero; I was going to stay there. So I arrived there at 8:00 in the morning and I didn't call for help till midnight. And by the time help arrived, I was pretty well exhausted and that was wrong. It's not good. You need to have that rest. If you're going to save some mine rescue men that you go on a 6-hour or a 12-hour cycle, then you better do the same thing.

SPEAKER: Don't stay as long as we used to because our agency's taken a pretty strong stand on people not staying too long at the site because of the probability of making bad decisions after being on the site, being tired. When we, in District 6 when I was over there, when we first formulated our plan, the decision was made at that time at the local level to not stay more than 14 hours and this was under certain conditions. For example, if this occurred during the shift that I was working then and I went to the site, then I needed to realize that 14 hours would be my max, that there's a good probability that I'd make bad decisions after that. Since then we don't stay that long. We try to change out each shift, and our response plan would indicate this. So far as the people at the site doing the work, we would want to change that out to each shift. Better decisions.

The key insight to take from the discussions of time is that when time is better managed, the responders will have a better chance of making good decisions.

In addition to time as a stressor, responders have problems with information gathering and task sharing. This is especially true if they arrive on the scene at a point when first responders have, or should have, their temporary organization well underway.

SPEAKER: If I were to use our McClure Mine explosion as an example, it was as I would have expected it to be if the plan were in operation, and the plan was in operation. And it was very efficient organization where you set up and move it, and I'm driving an hour and a half. That was in the middle of the night.

I have gone to them in years past when it was absolute chaos, where there is no organization whatsoever. Mine management had little or no knowledge on the things that needed to be done, and there is chaos, total disorganization.

Generally, when I was with the government and I went to a site — the people that brought me up to date, it was done by one of our own people, one of the Bureau people. Obviously, they would know how many people are involved. They generally have a fairly decent idea of their location underground. In most cases — in one we did not. They'd have an idea of the quality of air in the mine, or at least coming from the mine, and have a halfway decent guess of our opportunities for getting back in there and getting some recovery efforts done.

I guess the only one I was involved in when no one really knew where people were was an inundation at a place called Holly Falls in West Virginia back in the middle '60s. They cut into an old mine, which flooded this mine, and really no one knew where the people were. There just wasn't any knowledge at all; they just had no idea. They knew where they were supposed to have been. But it was an interesting situation — for several days. It was done very well, and they just began the process of pumping and laying pipe, and moving pumps down, laying pipe, moving pumps down, and kept pumping the water down and also drilled some holes in locations where they thought people might be. Trying — based upon elevations in the mine. There weren't any successes as far as finding anybody through the hole, but eventually they got the water pumped down and broke the water seal, and people began to — they were able to find people. Sitting on a little bit of high ground, and they were just pure lucky, but they did find, if I remember, there may have been about 10 survivors and maybe 6 drowned. I don't know, something like that.

SPEAKER: If I'm the first to arrive, I want to assume all roles. Then, as other individuals arrive, I think you should delegate those roles out. But like I said earlier, I think there are common roles, such as supplies, manpower allotment, feeding people, setting up the information office to notify the rest of the MSHA. You know, again, it depends on how, what the particular event is. So there certainly should be a structure set up, and we should train for that structure of organization.

SPEAKER: I think as I said earlier, I called it "organized confusion." You're all onto the property, and you walk in, and everybody's pouring over their maps. Somebody is, there's usually three or four guys on the telephones contacting the underground areas to find out where the men are. There's two or three men checking out the check-in board to make certain that the men all are accounted for. And it's my job, of course, to make certain that the team get their field test—to get ready. And it's to be facetious, say, "Panic City," but that's not true, that's not true. The fact that everybody is moving, and they know what they have to do. Some of them are a little slow. They've got to get their book out. "Hey, what am I supposed to do here?" And they go down the list, "Oh, yeah." And then, as soon as they know what they have to do, and then it gets done. Fortunately, we haven't been in a situation where a life has been at stake.

I was called to one incident, and the man had been exposed to smoke inhalation, and he was in the refuge station, and he was in a safe area. He had been exposed. He was coughing and choking a little bit. But the rest of the mine hadn't been accounted for yet. And we were slow to respond to this guy because he's in the refuge station. He's in fresh air. He can do nothing but get better. And we were hesitant to send the team after him, when there could have been the possibility of somebody else in a greater danger than he was. And if the man would have collapsed or went down, then, of course, it would have been a case to go down immediately. But it was well organized. And there was a decision there that we were just hesitant to go after this guy when we didn't know where everybody else was. I think that's when, again, repeating what we said earlier, one of the rules that we have. The first and foremost is, "Where are the men?" Nobody moves until we find out where the men are. If there's somebody missing, then we're going to go after them. The incident that I've already mentioned, that Frug fire, that was a situation where the fire in the early stage of the fire and the communications went out. So we couldn't contact anybody, and there were 40 men that you couldn't contact. And you could call my corner "Panic City," because there were three teams going underground. There were three approaches to that area, and there was a team going down each approach to get to that area. We had to find where those 40 men were. The fire was the least of our concern at this stage of the game, "Where are those men?" Then when we went down and found that it was just a lack of communications, and the men were safe in their [refuge] station—fine. Now, we'll go fight the fire, and we're very proud of our organization. There is the decisions that have to be made, but they're usually made quick—usually.

SPEAKER: Well, when you get there, usually you're covered up with spectators. And then, soon after that, you're covered up with the news media, and those people have to be controlled, and you have to set up to deal with the—to deal with those people, and also, the people that are underground.

I guess the first; it's kind of a confusion, and then after you get there and get things settled, then it settles down. But there's a time when it would be in any emergency, because you got to get people there.

Another way of stating the notions advanced above is to point out that a scene may be a "buzzing confusion" to a new arrival, but make perfect sense to someone who has been there for several hours.

The veterans were asked the following questions about the criticality of information: "Throughout a response, what information is needed? What kind of technical support is important to gather this information? Who provides this support?" A selection of their answers follows.

SPEAKER: It's all in the control center. One, we've had separate lines. Phones that are not even connected with other lines. It's just on the one-to-one basis pager phone. And everything is given either by communication, or the log is given in writing, also. Copies are issued out. So we keep people informed what's happening. And also, even then during that situation, to inform the employees where they're going on to that are not involved in the emergency. So they made them feel more comfortable, especially if it's people that they work with that are still underground. There's a way of informing the people in the family, et cetera.

It's a guess to get material needed, you know. Trying to, what all the situation brings in safety, where you'd be. Have to call to get the situation. It's not here in town. You got to call out of town, so you got to try to get it here the fastest way the, air, et cetera. So, that is a little bit frustration of waiting. But that's the situation here, you know.

SPEAKER: I'll explain it, and it takes some explanation. It happened right at the beginning of a mine fire. The people that were at the mine fire, that were fighting it, had all their energies focused on putting the fire out. You had people in the command center that were able to analyze all this other information. Fan samples, analysis of air, all the other things that were taking place around putting a hose on the fire. And these outside people, the command center people, knew from this other information, that an unsafe condition was being created. So they ordered this group to come out of the mine. Well, the group did not want to come out. They almost flatly refused to leave, 'cause they said if we leave, we've lost our jobs; they close the mine. Well, it wasn't the mine rescue team, it was just a group of miners and foremen whose livelihoods were dependent, directly dependent. They weren't concerned with life or death, they were concerned with livelihoods of their families and stuff. The decision to evacuate was made by a senior MSHA person. I saw him put a halt, and just stop, like mid-sentence. And he covered the phone up and he turned around to the senior company official, and he said, "I'm telling you, those people are potentially going to get killed." Now, and he explained the situation to them, and then this senior official, who these men trusted, got on the phone and said, "It's not worth it. Come out." And they came out.

I watched that event and I greatly admired [the MSHA official], because he had the forethought to know that those people down there didn't trust his decision-making. What he had to do was put someone in communication with them that they trusted. And [he] didn't have a lot of time to waste, and that's what makes things successful. And that's why somewhere in this whole process you have to build this trust 'cause you don't have a lot of time to have a vote about it.

SPEAKER: I'll give you my procedure and what I teach the other representatives to do when they get there. Report to the health and safety committee chairman—the chairman of the health and safety committee, which is a local union officer. Then, immediately report to the command center to let them know that you're there, and get a briefing on the map, what you've got, because basically, the senior person there from the union is automatically in charge where you assume it or not. I mean, it's life, so I teach them to get a briefing from the union people. [Then], go to the company and get a briefing, and go to the command center and have somebody, either in the command center or off to the side, normally off to the side if you can work it out, to explain to you what you've got. And my procedure is that, when I get there, I try to avoid making any decisions or taking charge of our end of the operation for an hour or two until I can get the feel and know what everybody else is doing, then roll from there. 'Cause you go in there and try to initiate your own ideas, you don't know what's just happened, and so forth. It confuses everybody. That way you're not trying to find out what's happening and make decisions at the same time. So I usually try to take at least a couple hours before I tell the company, "Hey, from now` on, I'm in charge." You know, "If you want something out of the union, come to me." And up to that point, while I let the safety committee, or whoever's doing that there, I still let them do that, and I'll say, "I'm here to guide you."

Typically, I ask them what's going on, the status of the miners, if anybody's trapped or whatever, is anybody missing. And then I'll ask if they got a map. Never enough maps—first thing. You're always that first time is hunting for maps, and that's the most essential thing that you need on the spot, and then everything depends on circumstances from there. Once you get the map, you can look and see how the mines are laid out, how it's been ventilated, all that stuff. And then from that point, my next move is to find out what my air readings are. I want to know what my methane and everything's at. And then from there, you want to find out carbon monoxide. Then from there, it's just a big puzzle. You got to start putting it together—conditions, coal height, you know. Are you dealing in crawling coal, walking coal, wet conditions, dry conditions? That's a big part of it.

As the above accounts suggest, there are a lot of simultaneous tasks conducted at an emergency event. The better these tasks (and their timing) are understood, the better equipped people will be to engage in an effective response.

There is, unfortunately, limited theoretical or empirically derived knowledge about how people actually carry out their activities in reference to the elements prevalent at emergency worksites. Yet, a breakdown in any of the elements of an event can result in worker injury, stress, or other health problems. Thus, practitioners wishing to develop safety and health interventions find a limited stock of useful information to draw upon. This is linked partially to a paradigmatic gap that has led researchers to ignore the fact that a response scene is really an organized, though short-lived workplace. An emergency workday passes within an organizational structure that is transitory in nature because of the dynamic, evolving quality of an event. As a result, studies of completed emergency responses often fail to perceive patterns in the behavior and practices that occur in the course of an emergency workday. If work behaviors and practices are examined holistically, however, examples of continuity can be found. Thus, the key to effective interventions may be revealed in the individual experiences of emergency response participants, as well as in their shared nonmaterial culture.

References

Mitchell D [1990]. Mine fires: prevention, detection, fighting. Chicago, IL: Maclean Hunter Publishing Co., p. 1.

Salas E, Driskell JE, Hughes S [1996]. Introduction: The study of stress and human performance. In: Driskell JE, Salas E, eds. Stress and human performance. Mahwah, NJ: Lawrence Erlbaum Associates, p. 6.

CHAPTER 5.—SHARING LESSONS LEARNED

During the interviews, the emergency response experts were asked to discuss lessons they had learned through experience. The interviewers asked them to tell what they had learned that would cause them to handle similar situations differently and to tell about things they saw at past events that they would warn others not to do in the future. In response, as the previous chapters show, the experts discussed a variety of things. Most of their responses, however, touched on some of the same topics, including preparedness, experience, people on-site, mine rescue teams, and decision-making. A summary of their responses provides an overview of the lessons learned on-site at the largest mine disasters in the country.

Issue	No. of interviewees
Preparedness	9
Experience	5
People on-site	4
Mine rescue teams	4
Decision-making	3

Preparedness

Issues of preparedness were discussed in chapter 2, but will be revisited here briefly. Almost one-third of the interviewees suggested that future responders would be better able to handle emergencies if appropriate preparations have been made at the mine site. These experts argued that preparation begins with a viable plan that is practiced repeatedly.

There are two types of resources covered in a workable plan, each of which may have more than one element. First is personnel. The interviews pointed out a need for top-to-bottom staffing[1] with, secondarily, role definitions for those occupying the various positions: "Everybody should have a clear-cut understanding of what their responsibilities are, what their role is, and where they fit into the emergency structure." A number of veterans also recognized the advisability of setting up work schedules in order to keep workers reasonably well rested. A second type of resource is material. Time spent making sure needed supplies are available is time saved during an emergency: "'Hey, I'm at the mine. You guys are on standby. Get this stuff ready. We may need it.' It doesn't necessarily have to come to the property, unless there's a good chance you're going to use it, but at least have it ready."

---

[1] As much attention should be paid to providing for clerical support, for instance, as to providing for emergency managers.

Experience

All of the experts provided glimpses of what it was like to be on-site during responses to major mine emergencies. When asked what they had learned that should be passed on to future generations, five of them discussed how experience influences the effectiveness of responders. One veteran used the example of an emergency operation with which he had been involved early in his career. He concluded that what had taken him 30 hours to complete then would require only half the time now because of his experience. The five individuals also suggested ways to make the most of the learning opportunities that responses create. They talked about the hands-on learning that occurs during a response, about the value of reviewing events and sharing what can be learned from them, and about simulating emergency conditions to give trainees a preview of what they may encounter.

The five veterans recounted that most of their training occurred during responses. One individual discussed how learning can take place under these circumstances. He was not a decision-maker in his first few responses. Instead, he talked with the experienced person who was in charge. This experienced responder explained not only what should be done, but also the technical information that supported each decision. The novice workers asked questions throughout the response and gained invaluable knowledge from the seasoned "teacher." The interviewee believed that some aspects of mine emergency response have to be learned on-site through experience. He further pointed out that while presence during responses provides opportunities for learning, it is up to the individuals involved to ensure that teaching and learning take place.

Those who stressed experience during their interviews also thought that learning could and should take place through the review of events. One respondent related a story about reviewing an event to determine causation. He explained how pieces of the puzzle didn't fit together until well after the event when all testimonies had been reviewed and a report was being written. When this collected information was put together, an analysis became possible and recommendations could be developed to prevent a similar situation from occurring in the future. The significance of summary reports, or formalized hindsight, was also mentioned in terms of the importance of sharing these documents with others in the industry: "What we do is, we send [our association] a report, which they can take then and pass it on to other areas. Where, by our misfortune, they can learn by it, too."

It was also suggested that experience can be gained through simulated emergencies. One expert argued that there should be available facilities in which mock emergencies can be staged, exposing trainees to something like a real event. Fire training would include responses to burns in controlled environments, for example. "You learn from your mistakes. Give them opportunities to make the mistakes where no one is going to get hurt. And then we'll have people [who've been trained], and this is what we need because this is not something you learn in books."

People On-Site

Lessons that had been learned regarding who should and should not be allowed on mine property during a response were passed on by four of the experts. They expressed concern about people who were not taking active roles in a response, but who added complexity to the situation by interfering and/or simply contributing to overcrowding by their presence. These four individuals also addressed the special needs of victims' family members.

Control of site access had been discussed in another portion of the interview, but the experts believed that this issue was important enough that future responders should learn from their experiences.

> SPEAKER: [There] were situations where people had, I guess, influence with the company that got into the area that wasn't really needed. I think that there should be a very strict policy on the number of people that come in. And I don't think that any of them that I know about really adhered to that policy rigidly enough. I think maybe that in all cases there'd been some people in the area that shouldn't have been there.

The concern of extra people hampering response efforts is one consideration. A related (and equally important) issue is the safety of these bystanders.

> SPEAKER: [M]aybe [when] there'd been a rock fall, there's a little too much chance taken around that rock fall. With too many people when you ought to have a little bit less number of people in the area trying to get the person cleared of the rock. I don't think you need 8 or 10 people around trying to clear the rock, when they were in each other's way. And if there's another fall, you just have that many more people killed.

Generally, it was agreed there should be as many people as needed to conduct an efficient and effective response on the mine site and no more. It was mentioned, however, that special provisions should be made for the family members of missing miners.

> SPEAKER: [With my experience] I would [now] know that family members are going to be there, and they are going to be very, very apprehensive. Someone with compassion needs to pay a lot of attention to family members and be able to brief them and to make sure

that their pastors, their religious leaders, whoever they may be, [are] aware of the situation and invite them to come and be with the family members.

One expert agreed with the need to be as supportive as possible with family members, but warned that it should be clearly established who is and is not considered "family." He noted that in one case, family friends who were allowed to accompany the family abused their access to mine property in an attempt to gain more information about the victims. It is not surprising that these friends were interested in obtaining as much information as possible, but their activities hampered the efforts of responders. None of the veterans described this type of problem with actual relatives of victims, and all believed that they should be given every consideration possible.

Mine Rescue Teams

Four experts had opinions about mine rescue teams that they believed should be communicated to future responders. One issue was the problem of response times. Because team members may be away from the mine when an event occurs or may be called to a mine other than the one where they work, time is required to assemble a team. One suggestion for dealing with this delay is to use a mixed team: "You [aren't] going to call 14 men and 14 of them be at home. If I didn't get as many men from one team as I wanted to, I took one or two from the other team." This person warned that, although response time is important, team members (and other responders) should not endanger themselves by driving to the response in an unsafe manner. "There's no need to cause some more injury to yourself or someone else just to get there 2 seconds or 2 minutes earlier."

Another important issue relating to mine rescue teams is communication between a team and the command center. It was argued that teams sometimes do not follow directions of the command center and that they sometimes do not report back appropriately.

> SPEAKER: If you let [the teams] go and not know what they're doing, or for them to just call back what they want to tell you, how are you going to make a decision on the surface? You'll make the wrong decision probably three-fourths of the time because you don't know the information. And if they don't tell you, there's no way to know.

According to the interviewees, mine rescue teams should be the eyes and hands of their command center, but this has not always happened during responses. As stated above, roles and responsibilities must be clarified for everyone involved in the response before an event occurs.

Decision-Making

One set of questions on the interview guide covered the area of decision-making. Three of the interviewees believed some aspect of this issue should also be brought up when speaking of lessons they would like to pass on. In all three cases, the focus was on interactions between responsible individuals in the command center. It was pointed out that interplay between multiple people is helpful: "It's best to have somebody that you can talk to because no one person [can always know the best thing to do]. They just don't make them that smart."

On the other hand, conflict between individuals in the command center can be a problem. One person related the story of a "skirmish" that occurred between representatives of regulatory agencies during a response. Another pointed out that, as stated above with regard to planning, command center personnel must know their roles: "I have no problem with the four agencies (company, Federal, State, and union), as long as they understand that it's the responsibility of the company to call the shots." This individual stated that the government agencies and the union should provide personnel to assist the company if it needs help and to discuss any decision that may create a hazard. He further reflected that when "a person at a coal company has the knowledge of rescue and recovery work, it makes the job easier, and I think you get along better. Where a person does not have the knowledge, you'll have to question him more: 'Why are you doing this?' And the plans generally change [as a result of your questions]." A decision-maker whose plans are questioned may want to remember this advice from one of the interviewees: "Quick decisions is often bad. Try to count to 10 anyhow before you make a decision. And a wise man changes his mind, and a fool never does."

Training Issues

This section reports the responses given by these experts when asked how they think people who may have to respond to a future mine emergency should be trained. The interviewees discussed how training should be conducted, who should be trained, and what topics should be included.

The emergency response veterans who mentioned specific methods they think should be used for training future responders discussed three types of training: mine emergency response development (MERD) exercises, mock disasters, and tabletop simulations. Nine of the veterans said that some form of interactive simulated emergency response training would be the most beneficial. Several other ideas were also discussed.

Three of the experts believed that MERD exercises should be used to train responders. Historically, MERDs have been fashioned to be approximately day-long role-play events designed to present a realistic mine emergency scenario in the classroom to personnel who may be responding to an event. The overall goal of these exercises is to teach participants how to respond to a mine emergency in a correct, timely, well-organized manner that ensures the safety of all individuals involved in the emergency. At least one expert believed that every mine, regardless of its size, should conduct a MERD exercise.

> SPEAKER: I think we should require each mine, to have them put on their own MERD program. I think we could do it right at the mines. You know, we go to [classroom settings] and we set up mine offices, and mine foreman's office and stuff. Out there at the mines, it's the real thing. I think it would be more real and get more people involved.

Another form of enhanced mine emergency response training is the mock mine disaster. Like MERDs, mock disasters are role-play exercises designed to present a realistic mine emergency scenario. The major differences are that mock disasters make use of actual mine facilities and involve mine personnel who play their traditional roles at the operation. Three emergency response experts believed that conducting periodic mock disasters at mines would be an ideal way to train personnel. One veteran expressed his views on using mock disasters as training and assessment tools.

> SPEAKER: To me, training is number one. Continue training the people. Renew them again by having mock disasters. Bring in [consultants] that can actually just sit down and watch, and then come up with recommendations of what they've noticed during the emergency, what was supposed to have been done and was not done, where's our downfall, et cetera.

In addition to the role-play simulations discussed above, three response experts suggested that paper-and-pencil simulations of actual mine emergencies would be useful as training aids. Interviewees suggested that these simulations could be done in a classroom or in the offices at a mine site and used to train both command center personnel as well as other individuals who might be involved in the response. One expert described the success of using such simulations of actual mine emergencies to train personnel at his company.

> SPEAKER: All the little things that seem to be taken for granted, you run (personnel) through a couple of exercises and you suddenly find out, "Hey, I never thought of that." And [the exercise] brings all these little things to the front.

Continuing in his discussion about training with tabletop simulations based on actual mine emergencies that have occurred, this veteran said:

> SPEAKER: First, it was just hypothetical situations. And they didn't seem to prove too much 'cause they were too hypothetical, "Maybe this, maybe that." So then we just decided to take incidents—actual incidents that occurred and relive them. And it really brings things out, you know. "Are you prepared for this? Are you prepared for that?" When you speak of an actual situation in front of these guys, and they start to solve problems, then you can see—you see the panic starting to climb. You've got these men missing, and whatever you're doing is going wrong. You want to do this to ventilation, and the answer is, "No, that won't work, because the ventilation door is burnt up. You can't shut that door." "Oh, what are we going to do now? I can't shut the door." So it really proved it's working.

Finally, three response experts suggested a like number of alternative methods for training. One veteran responder proposed that operators have mine emergency response veterans come to the mine and make presentations on their past experiences, especially regarding critical phases of events. He suggested that mine operators could videotape the presentations and then show them to personnel who would be responding to emergencies. A second expert believed that training for future responders should begin early in their career, preferably at the college level, when students are receiving their formal training in mining engineering.

> SPEAKER: I strongly believe that's where it's [emergency response training] got to start is at the university, where you've got graduate students that are taking the mining engineering discipline. That [should be] a credited part of their required diploma, [where] they have to take emergency management training as part of that, so that when these engineers or whatever get into the workforce, they end up being mine superintendents, they've got that background.

A third response veteran suggested that a text-based document, providing recommendations for handling emergencies, be compiled and distributed to mine operators. This veteran believed that such a manuscript might be developed as a "how-to" manual and taken to the finest detail to ensure that all possible elements are covered. He believed that a document like this would be helpful, especially if written for use by any size or type of mine.

In sharing their thoughts on training future mine emergency responders, veterans mentioned specific groups of individuals they believed should be thoroughly trained in mine emergency response procedures. Their target audience included mine management, top (corporate) management, and regulatory personnel.

It is clear from their responses that veterans believe that both mine and corporate managers would benefit from training in mine emergency response. A 40-year veteran shared his thoughts on training mine management.

> SPEAKER: I would take the person that is responsible at the mine, and the mine foreman on each shift, and [they] would have mine emergency training continuously. I don't mean a shot of 7 hours, and that's it. I would [also] test [them] periodically.

In terms of training for corporate officials, another veteran responder said:

> SPEAKER: You ought to have a situation where you have primarily the top officers of the company involved, with regard to mine emergencies. Because when the emergency does happen, then they are the people that do get into that. I think they ought to be prepared more. And some of them that you have, probably never been involved in anything.

At least five veterans believe that both State and Federal regulatory personnel should be thoroughly trained in mine emergency response. One of these experts believed that not enough emergency response training is provided to mine inspectors.

> SPEAKER: We go to the academy[2] and we take too many senior people, and I think it needs to filter down to the individual inspector, 'cause any mine fire, that individual inspector is the person that's going to have to go out and really deal with [the emergency] in the initial quick stages of it.

Why is the issue of training mine emergency responders so important? The answer to this question can be found in one veteran's comment:

> SPEAKER: A lot of people's come and gone since 1969. And we're having less problems. So, in the next 10 to 15 years, there's just going to be a handful of people that's had any experience.

In short, fewer major mine emergencies are occurring. As seasoned emergency response personnel depart from the mining industry, there will be fewer individuals in the future who have first-hand experience in mine emergency response. If the suggestions made by mine emergency veterans for

---

[2] MSHA National Mine Health and Safety Academy.

training future responders are followed, the industry will be better prepared to handle emergencies that arise.

It is well known that some level of mine emergency response training is required by Federal and most State mine safety regulations. This mandated training is generally for front-line workers who often will be the first individuals to confront a mine emergency. Underground miners, for example, must receive training in escape procedures, first aid, and the use of emergency breathing apparatus and available fire-fighting equipment. Where mine operators have rescue teams, Federal and some State regulations define minimum training requirements for these teams. However, no regulations are known to exist that require comprehensive training for responding to and managing mine emergencies. Because there are no regulations in this area, the experts were asked what they believed decision-makers need to know to effectively manage a mine emergency. They identified five major subject areas: emergency response planning, mine ventilation, mine gas analysis, fire-fighting, and mine rescue.

Nine individuals suggested that responders be educated in overall emergency response planning. These experts believed this topic area should include training in the development of an emergency response plan followed by training of personnel in the implementation of the plan. Stressing the importance of training in mine emergency response planning, one veteran remarked:

> SPEAKER: They [mine management] should understand the requirement for preplanning. And they planned production ahead of time. They should plan for an emergency ahead of time. Because if that emergency occurs, all the planning in the world for your production, it doesn't mean anything. If you don't plan for the emergency, and the emergency occurs, you are not going to be able to handle it.

In another section of the interviews, experts were asked about issues that should be passed on to future responders. It is interesting to note that mine emergency veterans stressed the significance of preparedness with respect to (1) having a well-designed emergency response plan and (2) practicing for emergencies during that discussion as well.

Seven response experts stressed that responders should be given extensive training in mine ventilation, and five suggested training in mine gases. Many of these 12 individuals believed that most responders do not fully understand ventilation systems and mine gases well enough to make good decisions about how to proceed.

> SPEAKER: I think one of the most critical things that you encounter when you get to a mine is the ventilation system. You know what,

what has happened to the ventilation system, and what you need to do to restore the ventilation and, unfortunately, too many of us just don't know that much about ventilation, and I think that's one of the critical areas, whether it be an explosion, or a mine fire, or either one. That's one of the most critical areas to me, and we aren't that proficient in mine ventilation.

Another expert stressed the importance of emergency response personnel being well versed in ventilation.

SPEAKER: You've got to take care of your ventilation. If you're putting a lot of air in the mines and it's going over a fire, you're not doing anything. You'll never get it out. You may stop it [air] over here, but you may build up gas or make it a more dangerous situation over [there].

Finally, one veteran emphasized how knowledge of mine ventilation and mine gases can enhance the decision-making process.

SPEAKER: Training them, where we have people who can look at a mine map, understand ventilation, understand how air [behaves], [and] what the involvement of ventilation with the incident is, who understands what mine gases mean, who will be able to put things together and to make decisions that might alleviate this situation.

Veterans generally believed that future responders, especially the decision-makers in the command center, should be trained in, or at least have a strong understanding of, fire-fighting and/or mine rescue procedures. As suggested by several experts, all too often responders do not know how to properly attack a mine fire. Delay in figuring out how to handle a mine fire can result in the loss of valuable time and potentially the loss of the mine.

SPEAKER: Fire-fighting—how to fight a fire. A lot of people don't know how to fight the fire, particularly with the air, fighting a fire direct, and getting to it right away. You just can't stand around and wait till we call somebody to say, "We got a fire. What do you want us to do?" You got to get in and fight the damn thing. And you can't wait. Once you wait, particularly in the Pittsburgh Seam, you're a dead man. You're not going to have a mine.

Similarly, some response experts believe that decision-makers should be familiar with mine rescue procedures. As one veteran suggested, training of higher-level decision-makers in mine rescue procedures is important.

SPEAKER: I don't think that the operating supervision from the, shall we say, the nonproduction departments are aware of what the mine rescue personnel have to do and what types of situations that they get into. I think if they had a session on mine rescue [for the] department heads so [they] can have a better understanding of the terminology and the equipment that was being used, why it was being used, [and] how long it could be used.

Especially with emergencies at larger operations, individuals from regional company offices are often called in to help manage the event. Knowledge of mine rescue procedures can enhance the decision-making process.

CHAPTER 6.—SUMMARY AND CONCLUSIONS

This study has highlighted the major aspects of mine emergency response through the use of narrative accounts obtained from response veterans. These veteran responders were viewed as key people who had left, or were leaving, the mining industry. The methodology used in the interviews was to ask each person to explain how he came to be involved in response activities and then to center his stories upon specific moments in the progression of an underground mine emergency. As a result, each account followed certain themes that could be compared across situations: how I became involved in mine rescue, my first emergency response, what it's like to arrive at a disaster scene, my hardest decision, etc. It was found that much of the responder's cognitive efforts were spent in making sense out of the situation in which he found himself, whether it was his initial attraction to emergency work or his specific activities during an actual response.

The interviewees were also asked to discuss lessons they had learned through experience. The interviewers asked them to tell what they had learned that would cause them to handle similar situations differently and to tell about things they saw at past events that they would warn others not to do in the future. In response, the experts discussed a variety of things, but touched on some common topics, including preparedness, experience, people on-site, mine rescue teams, and decision-making.

When experts were asked how they would train future mine emergency responders, they answered in terms of how training should be conducted and who should be trained. Many believed that some form of interactive, simulated response training is the best method. Three types of simulation were discussed: mock disasters, MERD exercises, and tabletop simulations. Three major topics that should be included in training for emergency response decision-makers were identified by experts: future responders should be trained in mine emergency response planning and in the testing and revising of plans; future responders should be trained in mine ventilation and mine gases; and future responders should be trained in mine fire-fighting and mine rescue. Overall, devoting resources to training for a potential emergency was encouraged by all of these experienced responders.

Why has mine response veterans' knowledge been gathered and analyzed so that it can be provided to the next generation of miners in a concise and systematic manner? In short, fewer major mine emergencies are occurring. As seasoned emergency response personnel depart from the mining industry, there will be fewer individuals in the future who have first-hand experience in mine emergency response. Potential responders' experience will be obtained vicariously or not at all. If the wisdom of mine emergency veterans can be passed on and if their suggestions for training future responders are followed, the industry will be better prepared to handle the infrequent, but inevitable emergencies that arise.

APPENDIX A.—SURFACE ORGANIZATION OF UNDERGROUND MINE EMERGENCIES: INTERVIEW GUIDE

A. When I start the tape, I will ask you about your involvement with mine emergency response: when and why you became involved and things that stand out in your mind as you think back over your experiences.

Do you have any questions before we start?

[Tape on.]

1. Do you remember the first time that you witnessed an emergency situation at a mine? Can you tell me about it? Why were you there? What was your job at the time? Were you involved in the response?

2. Can you tell me about a response that you participated in that went particularly well? Why do you think that was so successful?

3. Can you tell me about a response that did not go so well? Why do you think those problems existed? How could it have been handled differently that may have been more successful? Why was that not done?

4. Was there anything that happened to you or that you saw during a response that you were not prepared for? What could someone do to prepare for that kind of situation?

5. What was the hardest situation that you found yourself in (or decision you had to make) during your work with mine emergency response?

[Tape Off.]

B. The next set of questions is about the decisions that are made during emergency responses.

Do you have any questions? Are you ready to continue?

[Tape on.]

6. Are there certain critical points common to many mine responses, times when certain actions should be done or decisions made?

What are the most important decisions that have to be made?

7. How are decisions usually made during an emergency response? Who makes these decisions? How many people (groups) are involved?

8. Can you give me an example of a response when the decision-making process worked well? What made the process work?

9. Do you remember any specific times that there were problems with decision-making during a response? What caused those problems? How did that affect the response? How could the situation have been handled differently?

10. Were you ever involved in a response when there was disagreement about what should be done? If yes, how was that dispute settled?

11. Was there a formal (written) response plan or procedure at any of the responses you attended? If yes, was it referred to during the response? Was the response different from responses that did not have formal plans? How?

12. Were you ever involved in a response where a consultant was on-site? Does that affect the decision-making? The relationships among the groups?

[Tape off.]

C. During the next section of the interview, I will ask for details about specific aspects of emergency responses. Whenever you can, please answer with examples from your experience.

Do you have any questions? Are you ready to continue?

[Tape on.]

13. How long after the discovery of a problem at a mine site were you usually contacted? Who contacted you?

14. How long after you were contacted did you usually arrive on the scene?

15. Can you tell me what it is like to arrive at a mine site during an emergency? What is happening?

16. Who is on-site during a mine emergency? Who should be? What groups are represented on-site? What is the role of each group? Do these roles ever change? When and why?

17. When you arrived at a mine site, did someone bring you up-to-date on what had already happened? If yes, who usually did that? What information is usually available at that point? Was information needed that was often not available?

18. Throughout a response, what information is needed? What kind of technical support is important to gather this information? Who provides this support?

19. Can you think of any situations when not having certain information caused problems? Can you think of a specific instance when having key information made an important difference?

20. How can availability of resources (equipment, supplies, people) affect a response? Can you tell me about a situation where needed resources were not on-site? How was that handled?

21. Since a number of responders can be involved for a fairly long period of time, how are the needs of those people taken care of? Can you give me an example of a time when those facilities were not adequate and one when this aspect of the response was handled well?

22. How important is a system for site security? Can you tell me of a time that site security affected a response?

23. How do media representatives at the scene affect the response? What are interactions with them like? When and how is information given to them?

24. Were you ever at a response when there were missing miners underground? How did that affect the response and the decisions that were made? Would things have been done differently if the miners were all accounted for?

25. How does the presence of family members or friends of missing miners affect the response? What are relations with these individuals like? How and when is information given to them?

26. Did you ever go underground during an emergency response? Why/why not?

27. How long did you usually stay on-site?

[Tape off.]

D. During this last set of questions, I would like you to think about all of your emergency response experience. I will be asking you about preparing future responders.

[Tape on.]

28. From your emergency response experience, can you tell me anything that you learned that would cause you to handle similar situations differently if you were involved in them in the future?

29. Can you think of anything that you saw done during a response that you would warn others not to do in the future? What and why?

30. Looking back over your career, were any important changes made in the way that mine emergencies were handled?

31. If you had all of the authority and resources you needed, how would you go about preparing members of the mining industry to manage mine emergency responses?

What would you recommend that is more practical/realistic?

32. What topics do you think should be covered in training for emergency response decision-makers?

33. Is there anything about emergency response that you would like to add?

Those are all the questions that I have. Thank you.

[Tape off.]

APPENDIX B.—GLOSSARY OF TERMS

Bleeder entries.—Panel entries driven on the perimeter of a block of coal that is being mined, and maintained as exhaust airways to remove methane from the working faces.

Bottom.—The floor in any underground mining cavity.

Brattice.—A temporary partition, usually covered with fire-resistant cloth, used in any mine passage to confine the air and force it into the working places.

Colliery.—Coal mine.

Command center.—A work area established in a mine office or other outside facility for the use of those directing an emergency response.

Crosscut.—A passage cut through a coal pillar to allow the ventilating current and mining equipment to pass from one entry to another.

Curtain.—A sheet of material used to direct air to the working faces.

District.—An administrative unit of the Mine Safety and Health Administration (MSHA).

Entry.—A coal heading or passageway about 20 feet wide and connected at intervals by crosscuts.

Fresh air base.—An underground station, located in the intake airway, which is used by rescue teams during underground rescue and recovery operations.

Gob.—An area of loose waste left in that part of the mine from which coal has been worked away.

Inby.—In a direction toward the working face, or interior of the mine, from any specific point indicated as the base or starting point. The opposite of outby.

Outby.—In a direction toward the surface from any specific point in the mine indicated as the base or starting point. The opposite of inby.

Panel.—A group of working places, usually operated as a unit, and separated from other working places by large pillars of coal.

Recovery.—The restoration of a mine or part of a mine that has been damaged by explosion, fire, water, or other cause to a working condition.

Rescue team.—A team of miners, from five to eight strong, trained in the use of breathing apparatus and in rescue and recovery operations.

Rib.—The side of a coal pillar or the wall of an entry.

Roof.—The rock, usually a shale, found immediately above a coal seam.

Seal.—To secure a mine opening against flowing or escaping gas, air, or liquids by injecting grout or building concrete barriers.

Shaft.—A vertical passage driven from the surface down through the strata to provide access to the mine workings.

Silo.—A tall tower, usually cylindrical and of reinforced concrete construction, in which bulk material is stored.

Slope.—An inclined passage driven from the surface down through the strata to provide access to the mine workings.

Stopping.—A wall, usually of concrete blocks and mortar, used to close off no longer needed crosscuts to prevent the air from short-circuiting and thus maintaining ventilation to the working faces.

Subdistrict.—An administrative unit within a Mine Safety and Health Administration district.

Ventilation arrangement.—The planned placement of a system of appliances, such as stoppings and curtains, needed to distribute air to a given location in the mine.

Working face.—The place at which work is being done in a heading or crosscut.

www.ingramcontent.com/pod-product-compliance
Lightning Source LLC
Chambersburg PA
CBHW071759170526
45167CB00003B/1091